THE *ESSEX* AND THE WHALE

THE *ESSEX* AND THE WHALE

Melville's Leviathan Library and the Birth of *Moby-Dick*

R. D. MADISON, EDITOR

PRAEGER™

An Imprint of ABC-CLIO, LLC
Santa Barbara, California • Denver, Colorado

Library of Congress Cataloging-in-Publication Data

Names: Madison, R. D., editor.
Title: The Essex and the whale : Melville's Leviathan library and the birth of Moby-Dick / R. D. Madison, editor [and introduction].
Description: Santa Barbara, California : ABC-CLIO, LLC, 2016. | Consists of excerpts from books read by Melville on whaling and whales with an introduction on his use of these sources to write Moby Dick. | Includes bibliographical references and index.
Identifiers: LCCN 2015046993 | ISBN 9781440850073 (hardback : alk. paper) | ISBN 9781440850080 (ebook)
Subjects: LCSH: Melville, Herman, 1819–1891. Moby Dick—Sources. | Whaling—Literary collections. | Melville, Herman, 1819–1891—Knowledge—Whaling. | Melville, Herman, 1819–1891—Books and reading. | Novelists, American—19th century—Biography. | Whalers (Persons)—United States—Biography. | Whaling—South Pacific Ocean—History—19th century.
Classification: LCC PS2384.M62 E87 2016 | DDC 813/.3—dc23
LC record available at http://lccn.loc.gov/2015046993

ISBN: 978-1-4408-5007-3
EISBN: 978-1-4408-5008-0

20 19 18 17 16 1 2 3 4 5

This book is also available on the World Wide Web as an eBook. Visit www.abc-clio.com for details.

Praeger
An Imprint of ABC-CLIO, LLC

ABC-CLIO, LLC
130 Cremona Drive, P.O. Box 1911
Santa Barbara, California 93116-1911

This book is printed on acid-free paper ∞

Manufactured in the United States of America

But it is a mild, mild wind, and mild looking sky.

CONTENTS

ACKNOWLEDGMENTS

Many people have contributed personally and directly to making this book possible.

Among those *qui ante nos fuerunt,* I give thanks: To Charles Wood, for introducing me to Murphy's *Oceanic Birds of South America.* To Harrison Hayford, for teaching me about book collecting and other things. To Howard Vincent, for pacing Hayford's backyard quoting "The Symphony." To Merton M. Sealts Jr. for establishing the system of keeping track of books physically associated with Melville. To John Seelye, for championing facsimile texts and material culture. To Wilson Heflin, for *two* copies of Maury's *Whale Chart.* To Don Yannella, for his voice and lunch in Annapolis. To Jonathan Wordsworth, for trusting me to row with him.

Among the living: To the Interlibrary Loan staff at the University of Arkansas, Fayetteville, for obtaining the right documents time after time. To Dean S. Nordquist, for introducing me to Nasrudin and Biblical scholarship. To Stuart M. Frank, for hiring me to sing chanteys at Mystic Seaport Museum. To Mary K. Bercaw Edwards, for systematizing the study of Melville's sources. To Ginger Vaughan, for describing Staunton's Blackfriars Theatre one day in the Folger Library reading room. To Steven Olsen-Smith, for taking over Sealts's work and creating the magnificent *Melville's Marginalia Online.* To Hershel Parker and Tom Tanselle, for "making fast" to the final Northwestern-Newberry volume. To Bob Sandberg, for transcribing

Melville's miscellaneous poems according to Tom's matrix—and thus freeing me from that task. To Michael Millman, for being my editor and friend.

And to Karen Lentz Madison, for she is of the tribe of Tiger, for the Cherub Cat is a term of the Angel Tiger.

Gramercy. *Vivant!*

INTRODUCTION: SWIMMING THROUGH LIBRARIES[1]

When he finished his manuscript of *Moby-Dick or the Whale* in the autumn of 1851, Herman Melville (1819–1891) may well have known more about whaling than any other person alive. "I have swam through libraries," Melville famously wrote in "Chapter 32: Cetology." In fact, Melville had been interested in whaling books nearly as long as he had been a whaler. He had first gone to sea as a merchant sailor in 1839, as his older brother struggled to keep the family solvent (their father had died in 1832). After a single voyage to Liverpool and back, Melville taught school briefly in New York and traveled to Illinois in search of employment: both endeavors were unsuccessful. Late in the year 1840, he made his way to New Bedford, the center of the American whaling industry. He sailed from Fairhaven, Massachusetts, in the brand-new whale-ship *Acushnet* on the third of January 1841 but jumped ship at the Marquesas Islands a year and a half later. He next signed on to the Australian whaler *Lucy Ann*, of Sydney, but was put ashore at Tahiti as a mutineer. After the *Lucy Ann* sailed without him, he signed on the whaler *Charles and Henry* of Nantucket for the passage to Hawaii, where the vessel arrived in May 1843. Three months later, he joined the Frigate *United States* homeward bound and arrived at Boston on October 3, 1844.

Somewhere in the South Pacific, Melville was handed a copy of Owen Chase's *Narrative of the Most Extraordinary and Distressing Shipwreck of the Whale-Ship Essex of Nantucket.* Somewhat bulkier than a pamphlet, Chase's book was not exactly the kind of thing that would draw one to sea. By the time he encountered Chase's book, Melville, who had lived through the stormy North Atlantic run between New York and Liverpool, had also experienced a Cape Horn passage and the remote and otherworldly Galapagos Islands. As a foremast hand, Melville would have heard the water rushing past a few inches from his head every time he turned in; he would have experienced broken spars and blown-out sails, being sent aloft himself to make the tatters fast. He may already have experienced a shattered whale-boat or lingering disease. The whale fishery—or any place at sea, for that matter—was not a particularly safe place to pursue one's profession. But Melville experienced nothing like what happened to the crew of the *Essex.* Chase's book, in brief, told a tale that begins in the remotest part of the Pacific Ocean with the sinking of the *Essex* by a whale and ends with a gruesome description of survival cannibalism after thousands of miles of open-boat sailing.

In 1819, the whale-ship *Essex*, Captain George Pollard (1791–1870), left Nantucket, the famous whaling island off the coast of Cape Cod in Massachusetts. The *Essex* was bound on a whaling voyage to the Pacific Ocean, which meant the vessel had to sail halfway round the world just to reach the whaling grounds. She carried the normal complement of whale-boats, 30-foot double-ended boats that could be rowed, paddled, or sailed. Each boat carried a crew of four oarsmen in addition to the officer who commanded the boat and a harpooner who was responsible for initially striking the whale and making the boat fast by means of a harpoon and about a quarter of a mile of line.

At about the longitude of San Diego but nearly on the equator, after the *Essex* had been 15 months at sea, the boats were lowered to pursue a "pod" or group of whales. The pursuers, however, became the pursued when a large sperm whale—with apparent deliberation—rammed the *Essex* twice. Since the ship was made of wood and was carrying oil, it did not sink immediately, but quickly took on enough water to bring the upper deck to the waterline. Repair of the ship was hopeless, and after salvaging what they could from the ship, the crew was divided among the three seaworthy whale-boats. When the *Essex* began to break up from the motion of the water, the voyage of survival began.

There is no land in the part of the ocean where the *Essex* went down. Although the men were initially well provided with food and water in their boats, they still had to make their way across thousands of miles of open water to safety. The men decided *not* to attempt to reach the Marquesas Islands, the nearest land, to the southwest: they feared cannibals. The three boats were, however, successful in reaching Henderson Island, part of the Pitcairn group (they thought they had landed on the nearby island of Ducie). It did not appear that there was sufficient water to support the entire crew on Henderson, so a few members of the crew stayed behind (and eventually were rescued), but the three boats set off for the coast of South America, about 3,000 miles away.

Three months after the sinking, Owen Chase's boat was picked up near the coast of South America. Only three men were left: ironically, they themselves had had to resort to cannibalism to survive. Captain Pollard's boat was also picked up near the coast: only he and one other survived, and they also were forced to resort to cannibalism. The third boat was never found. This was the story Melville heard at sea, first probably as a legend, second through meeting relatives of the mate and reading Chase's book.

Melville's recollections of his first encounter with Owen Chase, Captain Pollard, and the *Essex*—as he wrote them down later in his own copy of Chase's book—do not reflect his own fears and horrors as a whaleman in the 1840s: written only after the bulk of *Moby-Dick* was drafted in 1851 or later, Melville's marginalia record a developing fixation on Captain Pollard as a remarkable man, a survivor like Captain James Riley (1777–1840) of the Brig *Commerce*, who survived shipwreck and African captivity in 1815 before returning to New York to inspire a young Melville who would later recall him in *Redburn* (1849). As in the later case of Chase, in *Redburn,* Melville presents his narrator as having actually seen Riley himself, the sufferer and survivor, and the face-to-face witness has made a greater impression than Riley's 1817 book *Authentic Narrative* could by itself. What Melville must have taken away from his *initial* encounter with Chase, however, was not so much the impression of a singular man as the impression of a sentient whale, one that could add to all the other dangers of the sea, the possibility of *deliberately staving in and sinking an entire ship.*

Melville did not complete the voyage of the *Acushnet*—known now only through an "abstract log" collected at the behest of physical geographer and navy lieutenant Mathew F. Maury, whose collected data underpin the science and logic of Ahab's quest in *Moby-Dick.* Maury provided every seagoing vessel he could with a blank form on which the captain could record wind and currents on a daily basis. If the vessel was a whaler, it would record whales seen as well. From the results, Maury created charts that showed the most efficient sea routes for any given season—and, likewise, the best times and places to find whales. Maury's preliminary *Whale Chart* (1851) may have been published too late for Melville's direct use in "The Chart," but it nevertheless pictorially represents data that derived in small part from the *actual voyage* Melville participated in. And since the original log of the *Acushnet* has not been located, the only record we have of Melville's first whaling voyage is the abstract the *Acushnet*'s owners provided to Maury, now in the National Archives.

The impetus for Melville to jump ship in the Marquesas Islands, as he did in 1842, may have had as much to do with the horror of imminent death as with the trials of degrading shipboard life or the prospect of the cannibalism resorted to by *Essex* survivors adrift in their whale-boats. Melville clearly did not escape the threat of cannibalism by deserting on Nukahiva, the largest island, and living, as he supposed, among the fierce Typees: the Marquesas were famous for their reports of cannibalism.

Melville did not seem to retain a strong fascination with the whale-ship *Essex* after he returned to America in the frigate *United States* in 1844. *Typee* (1846), his first book, certainly shows a fixation with cannibalism, but of the more ritual than the survival kind. When he needed to fill out detail of this fictionalized account of a sojourn on the Marquesas, Melville turned to—among other works—David Porter's *Journal of a Cruise* (1815), which, along with its Marquesan encounters, recorded incidents of a cruise in search of British whalers in the South Seas. Porter may have provided Melville with the most complete *alter ego* for his whaling years, and Melville plundered Porter especially in *Typee* and "The Encantadas" (1854). With *Typee,* Melville gained instant international fame, but some, including his publisher John Murray, doubted (with good reason) the book's autobiographical veracity.

As an established sailor-author in the circle of Evert Duyckinck and his brother George (1816–1878 and 1828–1863) in New York, Melville began a brief association with their journal *The Literary World* primarily as a reviewer of maritime works—among them J. Ross Browne's *Etchings of a Whaling Cruise* (1847). Through his association with the Duyckincks, who were supporters of the "Young America" movement in asserting literary independence from the Old World, Melville began to read more widely and deeply.

The review of *Etchings* captures a moment in Melville's development when he was ready to leave whaling behind. Melville equated Browne's book with famed sailor-Brahmin Richard Henry Dana Jr.'s *Two Years before the Mast* (1840), both capstones to their respective professions of merchant sailor and whaler. The hardships Melville finds in Browne are serious and many, but he appears not to have Chase and the *Essex* on his mind as he recalls those hardships in his review. "Enough has been said," Melville writes, "to convince the uninitiated what sort of a vocation whaling in truth is." In 1847, at least,

Melville did not see a need for another book on whaling. And at that moment, to Melville, the horrors of the fishery did not include the ultimate fate of the *Essex*, sunk by a whale.

The horrors Melville may have recalled were not simply literary. At the time Melville was writing his review, his younger brother Thomas (1830–1884) was somewhere at sea on the whaler *Theophilus Chase,* having sailed in May 1846 to return two years later after an only modestly successful voyage. He had sailed, apparently, to avoid a female entanglement ashore. But between the allurements of the islands and the dangers of whaling, what could the family have been thinking when they let him go?

In 1847, Melville made overtures to meet Dana, and in July they met on the eve of Melville's marriage to Elizabeth Shaw (1822–1906). Dana recorded meeting "Hermann Melville (Typee)" with the parenthetical presumably referring as much to Melville's reputation as to the title of his book. Melville does not seem to have left his impression of Dana, although he can hardly have refrained from imagining his new acquaintance as beachcomber as well as a bastion of Boston society. Perhaps Melville saw Dana, at that time, as a specimen akin to Riley or Chase. It would be two years before they actually corresponded and then only on the occasion of a request for a letter of introduction.

Melville intended his second book, *Omoo* (1847), to continue the autobiographical strain of his first. Scholars concede that *Omoo* is less embellished than *Typee*, but Melville nevertheless drew an icy response when he proposed a third book to Murray in 1848. Melville toyed with the idea of another autobiographically grounded work, possibly even one that focused on the profession of whaling, but soon chafed at the narrow room such a work offered. He reported to Murray that he had wholeheartedly embraced romance and insisted that Murray at least take a look at the final product. Murray, firm to his practice, declined the work of pure fiction. Melville was exceedingly fortunate that the work was nevertheless immediately accepted, through the agency of a friend, by fiction-specialist Richard Bentley, and *Mardi* was published in the spring of 1849. *Omoo* had been completed by July 1846; *Mardi* by January 1849. Melville's next two books, *Redburn* (1849) and *White-Jacket* (1850) would be completed in about 12 and 10 weeks respectively.

The extraordinary time devoted to the composition of *Mardi* may reflect Melville's reconception of himself as an author, moving away from mere adventure to political and philosophical allegory. As in the case of his first two novels, *Mardi* reveals dependence on external sources, among them possibly Charles Knight's *Penny Cyclopædia of Useful Knowledge* (1833–1843), a set that Melville owned before the composition of *Redburn* and which he plundered heavily for *White-Jacket*, *Israel Potter* (mentioned by Melville in 1849 but not serialized until 1854–1855), and *Moby-Dick* (1851). *Mardi* also shows evidence of Melville's having before him Frederick Debell Bennett's *Narrative of a Whaling Voyage round the Globe* (1840), a work he may have seen as early as the composition of *Omoo* but which would become one of Melville's most important whaling sources for *Moby-Dick*—in part because of the work's elaborate appendices on the whale fishery (Browne's book also had an appendix on whaling, but its focus was on economics rather than material or social culture).

Melville's third book is downright perplexing in its recollection of the *Essex* disaster. *Mardi* is a dreamy political allegory that is not regarded as one of Melville's more successful books. But it does begin with an effectively described whale-boat voyage that might be expected to recall some of the details of the *Essex* account. In an elaborate discussion of sharks in Chapter 13, Melville actually refers to "the good craft Essex" as being "sunk by sea-monsters, as the alligator thrusts his horny snout through a Caribbean canoe." But when his own characters are becalmed only a few pages later in Chapter 16, the *Essex* and the fate of its crew are curiously absent from Melville's detailed description of dehydration in a whale-boat. The closest Melville comes to considering the unspeakable is his desire for solitude: "Like a desperado giving up the ghost, I desired to gasp by myself. . . . From being cast away with a brother, good God deliver me!" But the tension between the narrator and Jarl, the old salt with whom the narrator shares the boat, is over the dwindling supply of water—not any potential cannibalism that might be required in the aftermath.

If Melville badly gauged the reception of such a rhapsodic work as *Mardi*, he had nevertheless learned a great deal about the craft of writing. The extreme speed of composition of *Redburn* and *White-Jacket* is not indicated by botches in the text but, on the contrary, reveals the workings of an author who knew he would be buttressing his own experiences with external source material and who had at hand key sourcebooks, including the *Penny Cyclopædia*. One book in six months would have been quick work; two was astonishing. Melville's dismissal of first *Redburn* (a "little nursery tale," he wrote to Dana in October 1849) and later *White-Jacket* ("almost entirely for lucre," again to Dana, in May 1850) as insignificant works is belied by both their contemporary success and their legacy. With the two latest books, Melville was four-for-five since he took up his pen.

The years 1847 and 1848 had been pivotal for Melville, literarily as well as personally. He had learned, through his association with the Duyckinck circle and perhaps especially through his father-in-law Chief Justice Shaw's borrowing privileges at the Boston Athenaeum (a private library whose members included the most prominent Transcendentalists), to be a deep thinker. But he had not learned to become a deep writer, and he knew it. He knew exactly where his future security as a writer stood, and his deliberately executed work of 1849 shows that understanding. But toward the end of 1849, Melville must also have realized that he was running out of saltwater autobiography to underpin his fanciful reshapings. The unfruitful world of whaling had furnished him with the starting points of three books, the same number of whalers he had served on. *Redburn* had been focused on his first and only voyage as a merchant sailor, and *White-Jacket* had exhausted his experience in the navy.

In late 1849, Melville decided to attempt to peddle the recently completed *White-Jacket* directly to a London publisher. He saw this trip not only as a marketing ploy but also—if he were offered enough for his book—as a possible Grand Tour. As he prepared for the journey, he solicited letters of introduction through his family and friends. Not having actually developed a personal friendship with the Dana, Melville asked his father-in-law to approach this reigning maritime writer in America. Dana complied, and although his letter responding to Melville is now lost, we know from the content of Melville's October 6 response that in this letter Dana had suggested that Melville write up his man-of-war life in the Frigate *United States*. We do *not* know whether Dana had suggested such a thing previously or in what form he made the suggestion. We *do* know that Melville waited until the very last possible moment to thank Dana and to report that "A printed copy of the book is before me." This was not strictly true, since Melville had only proof sheets of *White-Jacket* destined for England at that point, and Melville's response may offer the slight taint of feeling patronized at having his literary course set by another. In writing a book about the navy—whether at the suggestion of Dana or in truth as the next logical step in his fictional biography—Melville must have known he was not really moving away from Dana's area of expertise. While Melville was at sea, Dana had publicly defended Alexander Slidell Mackenzie of the U.S. brig *Somers* for hanging an officer and two sailors for plotting mutiny. Melville's cousin, Guert Gansevoort, was second in command of the *Somers*. Melville was definitely not emerging from the shadow of Dana in showing the light and the dark of naval service. The powerful bond Melville felt on reading Dana's *Two Years before the Mast* was a tie with dangers as well as rewards. And yet Dana, who had "experienced in person the usages to which a sailor is subjected" and was now a perfectly sound member of Boston society, was theoretically the ideal reader for Melville's books. Melville, still rather a suspect character as "Typee," could benefit from a word from the now orthodox lawyer—and he did not hesitate to ask (half of his letter is devoted to the request) for that word publicly should *White-Jacket* encounter an unfriendly reception.

Most readers of the October 6 letter recognize a prefiguring of a trope Melville would often use as he tried to establish himself in brotherly love to another individual. Fictionally, the role of "sea-brother" began with Toby in *Typee* and reached its greatest expression in Jack Chase of *White-Jacket* (a real sailor, unrelated to Owen Chase of the *Essex*). In real life, Melville established (at least for his part) such a fraternal relationship briefly with Dana, somewhat longer with Nathaniel Hawthorne (1804–1864), and perhaps most fully with literary critic and art historian Henry T. Tuckerman (1813–1871).

In England, Melville disposed of *White-Jacket*, but not at the profit he expected. Melville indulged in a fortnight's abbreviated continental tour and then turned instead to sightseeing, literary acquaintance, and further book-buying in London. Several of the incidents of Melville's trip (corposants and men overboard, for instance) were to make their way into *Moby-Dick* in the ensuing two years, but Melville was signally uninterested in maritime literature—let alone whaling—while he was abroad. Of the books he obtained in London, only copies of his own *Mardi* and *Redburn* were maritime, while instead Melville seemed to be making up his mind about going forward with *Israel Potter*—a work that would not see publication for another four years. Of the books he *did* obtain in London, several would contribute not only to Melville's personal navigation around London in 1849 but also to the tone and setting of the contemplated *Israel Potter*. Charles Knight's *London*, for instance, had more to offer Melville the novelist than Melville the sightseer. Several gothic romances, including William Godwin's *Caleb Williams* (1794), invited Melville into a realm he would enter in the middle and late chapters of *Israel Potter*, and in *A Sentimental Journey* (1768; possibly in the volume in which Melville read *Tristram Shandy* [1759–1767]) Laurence Sterne provided Melville a Paris model for treating the risqué. Melville also found a London map from 1766: "I want to use it in case I serve up the Revolutionary narrative of the beggar," Melville recorded in his journal on December 18. Despite his recent reminder of the terrors of the deep, Melville's mind was not on whales.

Melville arrived home on the penultimate day of January 1850. He recorded no details of the passage home.

At this point in his biography of Melville, Hershel Parker devotes a chapter (33) to "The Breaching of Mocha Dick." His title does not refer to the *birth* of *Moby* Dick but to the recollection and inspiration of J. N. Reynolds's short story "Mocha Dick; Or the White Whale of the Pacific" (1839). Parker interprets Melville's marginalia in recently acquired books as a burgeoning of his fascination with cannibalism. Parker reports that "Melville had time," during the voyage home, "to conclude that he wanted to put off work on the Potter book." That he had the time is indisputable, but that he did so is speculation. We have nothing to go on in that direction. If Melville stayed in his cabin during the long, uneventful passage home, then he spent it contemplating the materials he had gathered for *Israel Potter*. There is no record of a whaling epiphany at sea, no record of an event that might prompt a mental return to the fate of the *Essex*.

To the length of the voyage, Parker ascribes the decision to turn away from *Israel Potter*. But even then, there was nothing to turn him toward whaling. He might just as well have been yearning for a rural bowl of milk after those long days at sea. Melville had not "shied away from" whaling. He believed there was nothing more to say and that he had covered, in his early books, whatever interesting scraps of poetry there was in whaling. In late winter of 1850, Melville was rereading James Fenimore Cooper, specifically Cooper's revolutionary narrative of the pirate, as one might call *The Red Rover*. Melville's review of the revised Putnam edition appeared in the *Literary World* in March.

In late spring—the exact date is not known—Dana again sent Melville a letter that included not only a favorable review of Melville's recent books but also yet another hint of the direction Melville's writing should take. Melville's response is complicated: Dana must have expressed some kind of brotherly sentiment to evoke Melville's gushing rejoinder:

> And I am specially delighted at the thought, that those strange, congenial feelings, with which after my first voyage, I for the first time read "Two Years Before the Mast", and while so engaged was, as it were, tied & welded to you by a sort of Siamese link of affectionate sympathy— ——that these feelings should be reciprocated by you, in your turn, and be called out by any White Jackets or Redburns of mine—this is indeed delightful to me. (May 1, 1850)

It is odd that Melville would go on to denigrate the very works that Dana had just praised, as written "by the job, as a woodsawyer saws wood." I take this as far more likely to be false modesty than the critical self-awareness most readers ascribe to the letter. "About the whaling voyage," Melville continues, "I am half way in the work, & am very glad that your suggestion so jumps with mine."

> It will be a strange sort of book, tho', I fear; blubber is blubber you know; tho' you may get oil out of it, the poetry runs as hard as sap from a frozen maple tree;—& to cook the thing up, one must needs throw in a little fancy, which from the nature of the thing, must be ungainly as the gambols of the whales themselves. Yet I mean to give the truth of the thing, spite of this.

Two days before, Melville had borrowed a two-volume set of William Scoresby's *An Account of the Arctic Regions, with a History and Description of the Northern Whale Fishery.* At about the same time, he ordered Thomas Beale's *The Natural History of the Sperm Whale*, which arrived from England by July 10—the nine weeks corresponding well with the time necessary for a packet to reach London with the order and for the book to be returned to New York.

For an experienced author with an established method for getting poetry out of a packet ship or a man-of-war, if Melville was ready to dive into *Moby-Dick* on the first of February, he took a singularly long time to invest in his major whaling sources. He had Browne and Bennett from previous encounters. He had his *Penny Cyclopædia* for everyday reference. Melville's letter to Dana expresses a methodology that, far from prefiguring *Moby-Dick* as we know it today, suggests much more that of his earlier works and of the subsequent *Israel Potter.* Such evidence as there is might more likely account for him being "half way" in the latter book—not as we know it today, after Melville's inevitable tinkering, but as a continuation of the kind of book Dana had expressed his appreciation of and that had initiated such a brotherly response.

The most important common denominator is the *Penny Cyclopædia of Useful Knowledge.* Melville had used, or would use, it to fill out detail in the John Paul Jones episodes of *Israel Potter.* The *Penny Cyclopædia* does not loom large in *Pierre* (1852) or other works apparently composed after *Moby-Dick* (although this certainly deserves more study). It would seem much more likely that Melville had decided in favor of continuing with the Revolutionary narrative of the beggar, and his claim to Dana that he was "half way" in a whaling work was the shallowest kind of cover after being nettled by a *second* nudge by Dana. The lady doth protest too much, methinks.

The Dana correspondence bears further scrutiny in another way, too. In noting Melville's presence as "Typee," Dana knew that Melville was the man who had lived among cannibals and, moreover, had gotten among them by running from a whaler. In his first letter—the one that suggested Melville write about man-of-war life—Dana was suggesting a course that was far from unique. Why on earth didn't he suggest that Melville write about whaling in the same letter? Why was he so guarded about sowing his germinous seeds?

At the earlier date, Dana may have thought, along with Melville, that the latter had covered the most interesting aspects of whaling already (generic dissatisfaction and jumping ship), and that in any case J. Ross Browne had specifically covered the whaling ground. What could have changed his mind?

While Melville was abroad in 1849 and early 1850, Rev. Henry T. Cheever, editor of the *New York Evangelist*, published *The Whale and His Captors*. Even though this work was advertised in the *Literary World*, the periodical for which Melville had been and would be again a reviewer, Melville somehow missed it. He used it only late in *Moby-Dick*, mainly in footnotes and other ancillary matter that did not require Melville to recopy pages. "This book," Cheever writes in the preface, "is simply what its title indicates, the mind of the author pretending to be only the camera obscura through which the rays from Nature and Nature's living scenes have passed uncolored to the canvas." The key word here is "pretending": far from being a photographic record of the voyage of the whaler *Preble*, on which Cheever sailed as a convalescent, the book is as stuffed with secondary matter as *Redburn* or *White-Jacket*. Cheever's book was published by Harper, Melville's own publisher at the time and the future publisher of *Moby-Dick*. It is simply incomprehensible that Melville stepped off the boat unaware of the initial release of a whaling book published by his own publisher and advertised in his "own" journal—unless he had no special awareness of whaling and hadn't been thinking at all about writing a whaling book.

No one was likely to point out the book to Melville: Cheever had reviewed *Typee* with some invective in *The Evangelist* in 1846. But Dana might well have noticed the book. That would account for his prompt to Melville in the spring of 1850: Dana would have recognized that Cheever's book, while renewing or feeding on interest in whaling, was not a voice from before the mast. Dana *would* have recognized the same allusion to Virgil's *Aeneid* that Dana had made in his own book: "Forsan et haec olim" (perhaps one day it will be pleasing to remember even these things). Dana may or may not have agreed with Cheever's assertion, "Neither does it enter into our purpose to portray a sailor's life and manners in the forecastle or before the mast, alow or aloft, for this is a department of marine literature in which books are so numerous, both in the form of the novel and the sea journal, that little remains to be told." Cheever uses the phrase "lights and shadows" twice in the succeeding passage, undoubtedly reminding Dana of his own words "the light and the dark together," from the introduction to *Two Years before the Mast.* Dana probably knew Henry's older brother George Barrell Cheever (1807–1890), at least by reputation: the latter Cheever was an outspoken proponent of the death penalty on the eve of the *Somers* affair.

Furthermore, on the eve of Dana's own voyage to the Pacific in 1834, the *North American Review* (founded in part and edited at one time by Dana's father and by now the leading literary journal in America) published the review article "The Whale Fishery," which listed at its head Scoresby's two major works, *The Arctic Regions* and *The Northern Whale Fishery,* and Harper's compilation *Discovery and Adventure in the Polar Seas and Regions.* This latter was the version of Scoresby known to, and cited by, Cheever—and as a volume in Harper's *Family Library* would have been readily accessible to Melville had he known of its existence in May 1850. In the article, Melville would have found yet another account of the *Essex* of the kind he referred to (in the annotations in his own copy of Chase) unsympathetically as "stupedly abbreviated." But Melville, unlike Dana, did not oscillate in the orbit of the *North American.*

The Whale and His Captors is not a bad book. It is rich in anecdote—some of it derived from the same sources Melville would use. Cheever had Beale's *London Quarterly* essay (not the 1839 book) and the Harper Scoresby, and probably other unidentified sources including, perhaps, the *North American* article. Cheever's illustrations (apparently woodcuts) are often derivative and run the gamut from monstrous to less erroneous—but seem to have escaped Melville's censure in *Moby-Dick*'s chapters on whaling iconography. Cheever's book is above all evangelical, with nearly every chapter including a pious twist at the end and concluding in two complete chapters on Sabbath-keeping. Melville missed it for a year and a half; Dana probably saw it when it came out and recognized that Melville could provide a better treatment, even along the lines of *Redburn* or *White-Jacket.*

Meanwhile, Melville approached his new English publisher, Richard Bentley, with a book proposal. On June 27, Melville wrote to announce that "In the latter part of the coming autumn I shall have ready a new work." It was now two months since Melville had written to Dana that he was "half way" in the work: Melville was usually cautious in proposing his books to British publishers; he seems to have waited until he had something substantial in hand. But Melville's letter to Bentley was something different: it proposed a book to be completed six months out. The previous summer, Melville had written *two* books. While he was in England, Melville clearly had no inkling he was going to write a whaling book: if he had, he would have proposed it to Bentley on the spot. Nor, after his long and supposedly epiphanic voyage back, did he immediately propose the new whaling book to Bentley. Instead (it has been assumed), he waited until two months after being "half way" in the book to propose a manuscript to be completed in another six. The dates just don't make sense.

Melville goes on to describe the work in mind:

> The book is a romance of adventure, founded upon certain wild legends in the Southern Sperm Whale Fisheries, and illustrated by the author's own personal experience, of two years & more, as a harpooneer.

Among the oddities of this statement (which doesn't hint at all about even the basic plot of *Moby-Dick*, a sinking inspired by the *Essex* story) is Melville's description of himself as a harpooneer. Most students of Melville have raised an eyebrow here because of the unlikelihood of Melville's having held such an important berth at all, let alone for two years. Did Melville ever personally strike a whale?

In all likelihood, there never was a narrative *Moby-Dick*, drafted before the arrival of Melville's key whaling sources. He had some of them—Bennett and Browne and *Penny Cyclopædia*—before he left for England. The last of these, with its emphasis on and praise of Beale and Scoresby, undoubtedly spurred Melville on to further book-collecting the moment he received Dana's letter. He began *Moby-Dick* with most of his cetological works in hand and the remaining one (Beale) on order.

Even after Melville reported his so-called progress to Dana, after he offered a work to Bentley to be ready by November or December, and after his friend Duyckinck reported at summer's end that the work was "mostly done," it would take another year to complete the work as we know it, transformed by Melville's reading of Thomas Carlyle and his personal acquaintance with Hawthorne.

Sometime, we don't know exactly when, Melville asked his father-in-law to track down a copy of Chase's *Narrative*, a book he had read once long ago and far away. He had, by the late spring of 1850, several outline accounts of the *Essex* tragedy. But at some point, Melville became fixed on the deeper tragedy of the figure of Captain Pollard. Reading Chase's book at sea had introduced Melville to a profound climax for a whaling book, and Melville encountered the *Essex* sinking again in several of his sources, but once he had embarked on his own literary whaling voyage Melville could not shake off this kraken of a man who had suffered the greatest of trials, who had survived by cannibalism. The "stupedly abbreviated" accounts focused on the terrors of the whale but glossed over the existential predicament of the men in the boats. Even Cheever was mostly interested in "the most dreadful display of the whale's strength," eschewing, as far as possible, "the last resort of hunger in the wretched, which words shudder to relate!" In fact, Cheever doesn't relate it. This is likely what Melville was looking for when he sought out

Chase's narrative—not wild legends of the whale fishery but a fuller account of living men who had looked long into the face of the fire.

In that sense, Melville's contemplation of Captain Pollard anticipates (or parallels) his exploration of psychological themes in his profoundly disturbing exploration of family guilt in *Pierre*, in addition to the philosophical problem of good and evil of the final version of *Moby-Dick*. Melville did not really turn away from anything when he began *Pierre*: he was following his leader.

Captain Pollard, considered among Melville's most genial friendships, represented an extreme of human experience that Melville could be astounded by but could neither fully understand nor approach. Sometime after his father-in-law finally located a copy of Chase's narrative, Melville recorded his current and subsequent impressions of the officers of the *Essex*. Chase might have been a candidate for soul mate: "He was a large, powerful well-made man," Melville wrote, erroneously mistaking a relative for the man himself, but nevertheless revealing the psychology of the encounter:

> rather tall; to all appearances something past forty-five or so; with a handsome face for a Yankee, & expressive of great uprightness & calm unostentatious courage. His whole appearance impressed me pleasantly. He was the most prepossessing-looking whalehunter I think I ever saw. [Melville'spelling regularized]

Melville's addition about Pollard is strikingly different:

> Since writing the foregoing I—sometime about 1850–3—saw Capt. Pollard on the island of Nantucket, and exchanged some words with him. To the islanders he was a nobody—to me, the most impressive man, tho' wholly unassuming even humble—that I ever encountered.

Significantly, Melville's copy—a defective one finally located by Judge Shaw's acquaintance Thomas Macy of Nantucket—ends with the crisis of cannibalism rather than the denouement of rescue. In a passage of his notes titled "Sequel" in his copy of Chase's book, Melville writes, "Captain Pollard's boat (from which Chase's had become separated) was also after a miserable time, picked up by a ship, but not until two of its crew had died delirious, & furnished food for the survivors." Melville reports that he had heard from a mate of the *Acushnet*, and therefore presumably at the time of his first encounter with the *Essex*, that "Pollard himself returned to Nantucket, & subsequently sailed on another whaling voyage to the Pacific, but he had not been in the

Pacific long, when one night, his ship went ashore on unknown rocks." Pollard, Melville reports, did not go to sea again.

Although Chase apparently had the misfortune to be cuckolded while at sea, it was Pollard who stuck in Melville's imagination. Decades later, when he was writing *Clarel: A Poem and Pilgrimage in the Holy Land* (1876), his long poem of belief and disbelief, Melville has his characters recall a figure Melville has created from the physical and moral traits of both Chase and Pollard, "one whom grim/Disaster made as meek as he/There plodding." Like his synthetic character in *Clarel,* Melville continued to brood on some secret thing.

R. D. Madison

Note

1. The working title for this introduction should be "*Moby-Dick:* Genetic Theories Reconsidered," as cited in the Northwestern-Newberry Edition of *Moby-Dick* in 1988. I was sure that Hershel Parker's massive research for the new *Log* and for his definitive *Herman Melville: A Biography* (1996, 2002) would turn up evidence of composition of *Moby-Dick* between January and May 1850. It would be an understatement to say that if Parker did not turn up new evidence, new evidence is not likely to be forthcoming. I hinted very strongly about my reading of the months leading up to Dana's second letter in "Literature of Exploration and the Sea" in Wyn Kelley's *A Companion to Herman Melville* (Malden, MA: Blackwell, 2006), especially 286–287, but the present introduction contains the discussion anticipated in the Northwestern-Newberry *Moby-Dick.*

When scholars talk about a "genetic theory," all they are saying is that they wonder—in great detail—how a book was written. We think of the Book of Genesis as the beginning of scripture or of genetics as the science of where we came from. Books may mean many things to different readers, but there is only one *true* story about how any book came to be written. Of course, that one true story may be very complicated itself—as it is in the case of Herman Melville's great American novel *Moby-Dick.*

As it turns out, most of the time we have only pieces of that true story—fragments of DNA, if you will. In the case of *Moby-Dick,* most readers recognize that Melville found inspiration for the climax of his book from the only widely known recorded sinking of a whale-ship—the one chronicled by First Mate Owen Chase in *The Narrative of the Most Extraordinary and Distressing Shipwreck of the Whale-Ship* Essex, the book published in 1821 and read by Melville some 20 years later while he himself was a whaleman in the South Seas.

Since 1947, when Charles Olson published *Call Me Ishmael,* no one who has looked into it believes that Melville simply sat down and wrote out *Moby-Dick.* Olson thought that there were two successive drafts of the book: the first without

Melville's Byronic Captain Ahab and the second with him. By 1954, George R. Stewart had developed a three-stage composition process, in which the early chapters came from an early conception of the story, the late chapters represented a new direction, and the middle chapters were revised to accomplish the transition. James Francis Barbour, in his 1970 dissertation and a 1975 article in *American Literature,* further complicated the three-stage theory. In 1978, on the basis of long hours spent with the manuscript of *Billy Budd* (the only complete surviving manuscript of a Melville novel), Harrison Hayford published *Unnecessary Duplicates: A Key to the Writing of* Moby-Dick, demonstrating that *Moby-Dick* shared the same evidence of the repeatedly revised passages at all stages of composition of *Billy.* After Hayford, readers confronted multiple *layers* of revision rather than *chapters.*

Today, Barbour's sequence of composition—when Melville was writing a story, when he was carrying on about whales, when he was writing a philosophical tragedy—is still mainly accepted, alongside Hayford's reminder that revision could and did strike anywhere.

Each of the scholars I've mentioned advanced the study of *Moby-Dick* in significant, dramatic, and positive ways. Rather than engaging any scholar systematically, my discussion above reviews the inception of *Moby-Dick* with only scattered references to other surmises about the origin and growth of the book. Needless to say, I am standing on the shoulders of giants.

Note on Sources

Quotations in this introduction and the individual headnotes come from works listed in the "Further Reading" chapter that follows and from the works anthologized in this book. Throughout, I have depended on Parker's *Herman Melville: A Biography* (1996, 2002) for biographical data, on the Northwestern-Newberry edition of *Correspondence* (Vol. 14, edited by Lynn Horth, 1993) for Melville's letters and the circumstances of their composition, and on the Northwestern-Newberry edition of *Journals* (Vol. 15, edited by Howard C. Horsford and Lynn Horth, 1989) for Melville's doings in England and France during his 1849–1850 trip.

FURTHER READING

Anderson, Charles Roberts. *Melville in the South Seas.* New York: Columbia University Press, 1939. General study of key sourcebooks throughout Melville's maritime *oeuvre.*

Bercaw [Edwards], Mary K. *Melville's Sources.* Evanston, IL: Northwestern University Press, 1987. The standard reference work in identifying all Melville's known and suspected sources.

Chase, Owen, Thomas Nickerson, et al. *The Loss of the Ship Essex, Sunk by a Whale.* New York: Penguin, 2000. Edited by Nathaniel and Thomas Philbrick, this work contains Chase's book along with other accounts Melville would not necessarily have seen but which might very well represent the "sailor talk" that he did hear at sea.

Frank, Stuart M. *Herman Melville's Picture Gallery.* Fairhaven, MA: Edward Lefkowicz, 1986. Absolutely indispensable guide to Melville's pictorial whales, profusely illustrated from many of the sources anthologized in the present collection.

Gaines, Kendra. "A Consideration of an Additional Source for Melville's *Moby-Dick.*" *Melville Society Extracts* 29 (January 1977), 6–12. The *Penny Cyclopædia* drew on many standard reference works that might otherwise appear to be Melville's sources.

Gidmark, Jill B[arnum], ed. *Encyclopedia of American Literature of the Sea and Great Lakes.* Westport, CT: Greenwood Publishing, 2001. Covers several of the minor writers represented herein.

Hayford, Harrison. "Two New Letters of Herman Melville." *ELH* 11:1 (March 1944), 76–83. The first discussion of the chronology of *Moby-Dick* to focus on letters that remain the earliest evidence for composition.

Hayford, Harrison. "Unnecessary Duplicates: A Key to the Writing of *Moby-Dick*." Faith Pullen, ed. *New Perspectives on Melville.* Edinburgh: Edinburgh University Press, 1978, 128–129. The most thorough consideration of the composition of *Moby-Dick.*

Hayford, Harrison, Hershel Parker, and G. Thomas Tanselle, eds. *The Writings of Herman Melville: The Northwestern-Newberry Edition.* Evanston, IL, and Chicago: Northwestern University Press and The Newberry Library, 1968–present.

Heffernan, Thomas Farel. *Stove by a Whale: Owen Chase and the* Essex. Middletown, CT: Wesleyan University Press, 1981. A thorough examination of Chase's book and its relation to Melville.

Heflin, Wilson. *Herman Melville's Whaling Years.* Mary K. Bercaw Edwards and Thomas Farel Heffernan, eds. Nashville: Vanderbilt University Press, 2004. By the discoverer of Melville's third whaler: the most thorough biographical examination of the voyages.

Melville, Herman. *Moby-Dick.* Hershel Parker and Harrison Hayford, eds. New York: Norton, 2002. 2nd edition. Contains the most easily consulted collection of analogues.

Olsen-Smith, Steven. "Melville's Copy of Thomas Beale's *The Natural History of the Sperm Whale* and the Composition of *Moby-Dick*," *Harvard Library Bulletin* 21:3 (Fall 2010), 1–77. Essential monograph on the most important whaling source.

Parker, Hershel. *Herman Melville: A Biography.* 2 vols. Baltimore: Johns Hopkins University Press, 1996 and 2002. By several magnitudes, the most thorough biography of Melville.

Philbrick, Nathaniel. *In the Heart of the Sea.* New York: Viking Penguin, 2000. General account of the *Essex* sinking informed by recent manuscript discoveries.

Philbrick, Thomas. *James Fenimore Cooper and the Development of American Sea Fiction.* Cambridge, MA: Harvard University Press, 1961. The definitive study of nineteenth-century American maritime writing.

Sanborn, Geoffrey. *The Sign of the Cannibal: Melville and the Making of a Postcolonial Reader.* Durham, NC: Duke University Press, 1998. Two key sections (indexed under "Shipwreck narratives") distinguish survival cannibalism from cultural perceptions of savagery.

Sealts, Merton M., Jr. *Melville's Reading: Revised and Enlarged Edition.* Columbia: University of South Carolina Press, 1988. Indispensable reference to books with a physical connection to Melville. See also *Melville's Marginalia Online.*

Vincent, Howard P. *The Trying-Out of* Moby-Dick. Boston: Houghton Mifflin, 1949. The most comprehensive study of the sources of *Moby-Dick,* still extremely valuable although dated.

A NOTE ON THE TEXTS

This anthology is meant to introduce readers to the literary side of Melville's whaling world and to whet their appetite for more detailed source study than has been accomplished to date. In cases where readers will want to explore individual texts further, the headnotes will refer them to the appropriate editions.

The selections have been made not only to illustrate Melville's direct borrowings, as they do in some cases and has been pointed out by many scholars, but also to suggest contributions in tone and resonance that might not be recognized through the necessarily brief excerpts that usually accompany source studies. Readers will recognize unstudied details like the side-ropes from Hart's 'Sconset and the loss of a leg from Scoresby's northern whale fishery. These are details that lingered in Melville's mind even when he was not concentrating on adapting a specific passage.

We know with certainty when Melville procured some books but know little or nothing about the acquisition of others. The selections in this anthology are arranged in a *plausible* order of Melville's encountering them, but a *definitive* order is impossible to determine. While I hope readers will find the present order useful, they should not take it as canonical.

It is likely, however, that Melville read every word of every text included here, and the books from which they came account for a good deal—perhaps most—of what Melville wrote about whales and whaling. They do not, of course, account for Ishmael and Ahab. In the summer of 1850, in meeting Nathaniel Hawthorne, Melville recognized that he himself was a literary "Catskill eagle." He felt himself diving and soaring with Shakespeare and John Milton, with Sterne and Carlyle, and with the prophets of the Old Testament. Whaling inspired the whale Moby Dick, but the book *Moby-Dick* evolved through Melville's unconquerable mind.

The texts in this compilation have been edited from a variety of sources. Not all of Melville's known whaling sources are represented here, but most are, while the length of the selection in each case represents my own bias and not scholarly consensus. The title of each entry identifies the edition most closely identified with Melville. The headnote provides identification numbers assigned by Merton M. Sealts Jr. in *Melville's Reading* (1988) and by Mary K. Bercaw (now Bercaw Edwards) in *Melville's Reading* (1987). The headnote further identifies, where appropriate, reliable modern editions such as could be purchased at a reasonable price or ordered through interlibrary loan. In the best cases, I have been able to transcribe my text from the original editions or excellent facsimiles. Regrettably, I have occasionally had to rely on poor facsimiles or other optically scanned texts as my starting point. I have done my best to restore these (often significantly defective) texts to their original readings and apologize for any errors that may linger.

ILLUSTRATIONS

With the exception of the cuts illustrating the texts of Beale and Colnett, the illustrations accompanying this anthology are reproduced from Cheever's *The Whale and His Captors* (1850). These illustrations are typical of those found in other popular whaling books, although some obviously have a great deal less authority. The titles here are those of Cheever's "List of Engravings"; they do not always agree with the captions. All *signed engravings are by W. Roberts; only one, the last, is signed by an artist, F. Purcell. The variety of formats suggests these were gleaned, at least in part, from earlier works.

For a full account of Melville's use of pictorial whaling sources, see Frank (1986).

A Polar Right Whale on the Ice.

There she blows! There she blows!
Man the boats! For nothing stay!
Such a prize we must not lose!
Lay to your oars! Away! away!

The Commodore Preble taking Supplies at Rimatara.

View of a Whale Ship in the Process of "Cutting in."

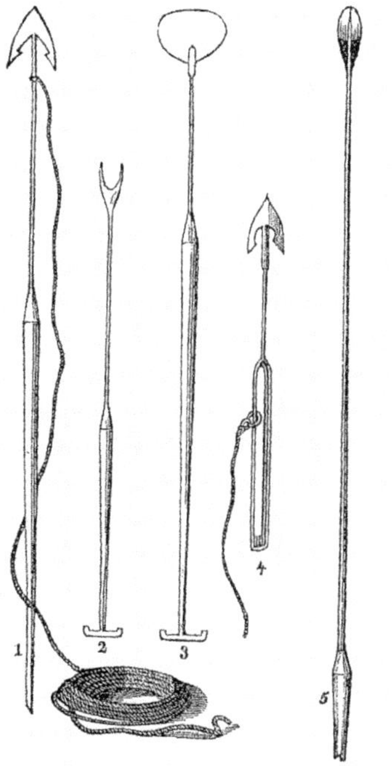

1. Hand Harpoon. 2. Pricker. 3. Blubber Spade.
4. Gun Harpoon. 5. Lance.

Launch the harpoon! Laugh at fear!
Plunge it deep, the barbed spear!
Strike the lance in swift career!

Give her line! Give her line!
Down she goes through the foaming brine.

He came up alongside of the Boat, and turned it over with his Nose as a Hog would his Eating-trough.

Combat between a Whale and the Sea Serpent.

View of a Whale-boat crushed by a Whale

Exploit of an American Harpooner in Delego Bay

Polar Whale dragging the Boats of the Bounty.

Picture of a Whale-boat thrown into a Whale's Mouth.

View of a Sabbath-keeping Whale Ship.

Picture of the Com. Preble Homeward Bound.

Herman Melville, "Chapter 45: The Affidavit," *Moby-Dick or the Whale*

(New York: Harper and Brothers, 1851)

INTRODUCTION

Moby-Dick was published in the autumn of 1851. Regarded as overwrought by its contemporaries, the book was not so well received as Melville's more straightforward narratives *Typee, Omoo, Redburn,* and *White-Jacket.* The book was, however, immediately recognized as belonging to a genre that included works by other mystical writers from the 17th to 19th centuries. *Moby-Dick* had a small appreciative readership from soon after its publication to the 1920s, when the book was rediscovered by both scholarly and general readers, and by mid-20th century, it became regarded as a classic of world literature.

Melville himself thought that several aspects of his novel would be regarded as incredible by the average reader, so he devoted "The Affidavit" to presenting basic facts about whales he had gleaned from his various whaling books, especially Owen Chase's *Narrative.* Without having to reread all of *Moby-Dick*, this chapter is a good place to anchor an exploration of those same whaling books. Modern Edition: Norton Critical Edition, 2nd ed., Parker and Hayford, eds. (New York: Norton, 2002).

PRIMARY SOURCE

So far as what there may be of a narrative in this book; and, indeed, as indirectly touching one or two very interesting and curious particulars in the habits of sperm whales, the foregoing chapter, in its earlier part, is as important a one as will be found in this volume; but the leading matter of it requires to be still further and more familiarly enlarged upon, in order to be adequately understood, and moreover to take away any incredulity which a profound ignorance of the entire subject may induce in some minds, as to the natural verity of the main points of this affair.

I care not to perform this part of my task methodically; but shall be content to produce the desired impression by separate citations of items, practically or reliably known to me as a whaleman; and from these citations, I take it—the conclusion aimed at will naturally follow of itself.

First: I have personally known three instances where a whale, after receiving a harpoon, has effected a complete escape; and, after an interval (in one instance of three years), has been again struck by the same hand, and slain; when the two irons, both marked by the same private cypher, have been taken from the body. In the instance where three years intervened between the flinging of the two harpoons; and I think it may have been something more than that; the man who darted them happening, in the interval, to go in a trading ship on a voyage to Africa, went ashore there, joined a discovery party, and penetrated far into the interior, where he travelled for a period of nearly two years, often endangered by serpents, savages, tigers, poisonous miasmas, with all the other common perils incident to wandering in the heart of unknown regions. Meanwhile, the whale he had struck must also have been on its travels; no doubt it had thrice circumnavigated the globe, brushing with its flanks all the coasts of Africa; but to no purpose. This man and this whale again came together, and the one vanquished the other. I say I, myself, have known three instances similar to this; that is in two of them I saw the whales struck; and, upon the second attack, saw the two irons with the respective marks cut in them, afterwards taken from the dead fish. In the three-year instance, it so fell out that I was in the boat both times, first and last, and the last time distinctly recognised a peculiar sort of huge mole under the whale's eye, which I had observed there three years previous. I say three years, but I am pretty sure it was more than that. Here are three instances, then, which I personally know the truth of; but I have heard of many other instances from persons whose veracity in the matter there is no good ground to impeach.

Secondly: It is well known in the Sperm Whale Fishery, however ignorant the world ashore may be of it, that there have been several memorable historical instances where a particular whale in the ocean has been at distant times and places popularly cognisable. Why such a whale became thus marked was not altogether and originally owing to his bodily peculiarities as distinguished from other whales; for however peculiar in that respect any chance whale may be, they soon put an end to his peculiarities by killing him, and boiling him down into a peculiarly valuable oil. No: the reason was this: that from the fatal experiences of the fishery there hung a terrible prestige of perilousness about such a whale as there did about Rinaldo Rinaldini, insomuch that most fishermen were content to recognise him by merely touching their tarpaulins when he would be discovered lounging by them on the sea, without seeking to cultivate a more intimate acquaintance. Like some poor devils ashore that happen to know an irascible great man, they make distant unobtrusive salutations to him in the street, lest if they pursued the acquaintance further, they might receive a summary thump for their presumption.

But not only did each of these famous whales enjoy great individual celebrity—Nay, you may call it an ocean-wide renown; not only was he famous in life and now is immortal in forecastle stories after death, but he was admitted into all the rights, privileges, and distinctions of a name; had as much a name indeed as Cambyses or Caesar. Was it not so, O Timor Tom! thou famed leviathan, scarred like an iceberg, who so long did'st lurk in the Oriental straits of that name, whose spout was oft seen from the palmy beach of Ombay? Was it not so, O New Zealand Jack! thou terror of all cruisers that crossed their wakes in the vicinity of the Tattoo Land? Was it not so, O Morquan! King of Japan, whose lofty jet they say at times assumed the semblance of a snow-white cross against the sky? Was it not so, O Don Miguel! thou Chilian whale, marked like an old tortoise with mystic hieroglyphics upon the back! In plain prose, here are four whales as well known to the students of Cetacean History as Marius or Sylla to the classic scholar.

But this is not all. New Zealand Tom and Don Miguel, after at various times creating great havoc among the boats of different vessels, were finally gone in quest of, systematically hunted out, chased and killed by valiant whaling captains, who heaved up their anchors with that express object as much in view, as in setting out through the Narragansett Woods, Captain Butler of old had it in his mind to capture that notorious murderous savage Annawon, the headmost warrior of the Indian King Philip.

I do not know where I can find a better place than just here, to make mention of one or two other things, which to me seem important, as in printed form establishing in all respects the reasonableness of the whole story of the White Whale, more especially the catastrophe. For this is one of those disheartening instances where truth requires full as much bolstering as error. So ignorant are most landsmen of some of the plainest and most palpable wonders of the world, that without some hints touching the plain facts, historical and otherwise, of the fishery, they might scout at Moby Dick as a monstrous fable, or still worse and more detestable, a hideous and intolerable allegory.

First: Though most men have some vague flitting ideas of the general perils of the grand fishery, yet they have nothing like a fixed, vivid conception of those perils, and the frequency with which they recur. One reason perhaps is, that not one in fifty of the actual disasters and deaths by casualties in the fishery, ever finds a public record at home, however transient and immediately forgotten that record. Do you suppose that that poor fellow there, who this moment perhaps caught by the whale-line off the coast of New Guinea, is being carried down to the bottom of the sea by the sounding leviathan—do you suppose that that poor fellow's name will appear in the newspaper obituary you will read to-morrow at your breakfast? No: because the mails are very irregular between here and New Guinea. In fact, did you ever hear what might be called regular news direct or indirect from New Guinea? Yet I tell you that upon one particular voyage which I made to the Pacific, among many others we spoke thirty different ships, every one of which had had a death by a whale, some of them more than one, and three that had each lost a boat's crew. For God's sake, be economical with your lamps and candles! not a gallon you burn, but at least one drop of man's blood was spilled for it.

Secondly: People ashore have indeed some indefinite idea that a whale is an enormous creature of enormous power; but I have ever found that when narrating to them some specific example of this two-fold enormousness, they have significantly complimented me upon my facetiousness; when, I declare upon my soul, I had no more idea of being facetious than Moses, when he wrote the history of the plagues of Egypt.

But fortunately the special point I here seek can be established upon testimony entirely independent of my own. That point is this: The Sperm Whale is in some cases sufficiently powerful, knowing, and judiciously malicious, as with direct aforethought to stave in, utterly destroy, and sink a large ship; and what is more, the Sperm Whale *has* done it.

First: In the year 1820 the ship Essex, Captain Pollard, of Nantucket, was cruising in the Pacific Ocean. One day she saw spouts, lowered her boats, and gave chase to a shoal of sperm whales. Ere long, several of the whales were wounded; when, suddenly, a very large whale escaping from the boats, issued from the shoal, and bore directly down upon the ship. Dashing his forehead against her hull, he so stove her in, that in less than "ten minutes" she settled down and fell over. Not a surviving plank of her has been seen since. After the severest exposure, part of the crew reached the land in their boats. Being returned home at last, Captain Pollard once more sailed for the Pacific in command of another ship, but the gods shipwrecked him again upon unknown rocks and breakers; for the second time his ship was utterly lost, and forthwith forswearing the sea, he has never tempted it since. At this day Captain Pollard is a resident of Nantucket. I have seen Owen Chace, who was chief mate of the Essex at the time of the tragedy; I have read his plain and faithful narrative; I have conversed with his son; and all this within a few miles of the scene of the catastrophe.[1]

Secondly: The ship Union, also of Nantucket, was in the year 1807 totally lost off the Azores by a similar onset, but the authentic particulars of this catastrophe I have never chanced to encounter, though from the whale hunters I have now and then heard casual allusions to it.

Thirdly: Some eighteen or twenty years ago Commodore J—, then commanding an American sloop-of-war of the first class, happened to be dining with a party of whaling captains, on board a Nantucket ship in the harbor of Oahu, Sandwich Islands. Conversation turning upon whales, the Commodore was pleased to be sceptical touching the amazing strength ascribed to them by the professional gentlemen present. He peremptorily denied for example, that any whale could so smite his stout sloop-of-war as to cause her to leak so much as a thimbleful. Very good; but there is more coming. Some weeks after, the Commodore set sail in this impregnable craft for Valparaiso. But he was stopped on the way by a portly sperm whale, that begged a few moments' confidential business with him. That business consisted in fetching the Commodore's craft such a thwack, that with all his pumps going he made straight for the nearest port to heave down and repair. I am not superstitious, but I consider the Commodore's interview with that whale as providential. Was not Saul of Tarsus converted from unbelief by a similar fright? I tell you, the sperm whale will stand no nonsense.

I will now refer you to Langsdorff's Voyages for a little circumstance in point, peculiarly interesting to the writer hereof. Langsdorff, you must know by the way, was attached to the Russian Admiral Krusenstern's famous Discovery Expedition in the beginning of the present century. Captain Langsdorff thus begins his seventeenth chapter:

"By the thirteenth of May our ship was ready to sail, and the next day we were out in the open sea, on our way to Ochotsh. The weather was very clear and fine, but so intolerably cold that we were obliged to keep on our fur clothing. For some days we had very little wind; it was not till the nineteenth that a brisk gale from the northwest sprang up. An uncommon large whale, the body of which was larger than the ship itself, lay almost at the surface of the water, but was not perceived by any one on board till the moment when the ship, which was in full sail, was almost upon him, so that it was impossible to prevent its striking against him. We were thus placed in the most imminent danger, as this gigantic creature, setting up its back, raised the ship three feet at least out of the water. The masts reeled, and the sails fell altogether, while we who were below all sprang instantly upon the deck, concluding that we had struck upon some rock; instead of this we saw the monster sailing off with the utmost gravity and solemnity. Captain D'Wolf applied immediately to the pumps to examine whether or not the vessel had received any damage from the shock, but we found that very happily it had escaped entirely uninjured."

Now, the Captain D'Wolf here alluded to as commanding the ship in question, is a New Englander, who, after a long life of unusual adventures as a sea-captain, this day resides in the village of Dorchester near Boston. I have the honour of being a nephew of his. I have particularly questioned him concerning this passage in Langsdorff. He substantiates every word. The ship, however, was by no means a large one: a Russian craft built on the Siberian coast, and purchased by my uncle after bartering away the vessel in which he sailed from home.

In that up and down manly book of old-fashioned adventure, so full, too, of honest wonders—the voyage of Lionel Wafer, one of ancient Dampier's old chums—I found a little matter set down so like that just quoted from Langsdorff, that I cannot forbear inserting it here for a corroborative example, if such be needed.

Lionel, it seems, was on his way to "John Ferdinando," as he calls the modern Juan Fernandes. "In our way thither," he says, "about four o'clock in the morning, when we were about one hundred and fifty leagues from the Main of America, our ship felt a terrible shock, which put our men in such consternation that they could hardly tell where they were or what to think; but every one began to prepare for death. And, indeed, the shock was so sudden and violent, that we took it for granted the ship had struck against a rock; but when the amazement was a little over, we cast the lead, and sounded, but found no ground. . . . The suddenness of the shock made the guns leap in their carriages, and several of the men were shaken out of their hammocks. Captain Davis, who lay with his head on a gun, was thrown out of his cabin!" Lionel then goes on to impute the shock to an earthquake, and seems to substantiate the imputation by stating that a great earthquake, somewhere about that time, did actually do great mischief along the Spanish land. But I should not much wonder if, in the darkness of that early hour of the morning, the shock was after all caused by an unseen whale vertically bumping the hull from beneath.

I might proceed with several more examples, one way or another known to me, of the great power and malice at times of the sperm whale. In more than one instance, he has been known, not only to chase the assailing boats back to their ships, but to pursue the ship itself, and long withstand all the lances hurled at him from its decks. The English ship Pusie Hall can tell a story on that head; and, as for his strength, let me say, that there have been examples where the lines attached to a running sperm whale have, in a calm, been transferred to the ship, and secured there; the whale towing her great hull through the water, as a horse walks off with a cart. Again, it is very often observed that, if the sperm whale, once struck, is allowed time to rally, he then acts, not so often with blind rage, as with wilful, deliberate designs of destruction to his pursuers; nor is it without conveying some eloquent indication of his character, that upon being attacked he will frequently open his mouth, and retain it in that dread expansion for several consecutive minutes. But I must be content with only one more and a concluding illustration; a remarkable and most significant one, by which you will not fail to see, that not only is the most marvellous event in this book corroborated by plain facts of the present day, but that these marvels (like all marvels) are mere repetitions of the ages; so that for the millionth time we say amen with Solomon—Verily there is nothing new under the sun.

In the sixth Christian century lived Procopius, a Christian magistrate of Constantinople, in the days when Justinian was Emperor and Belisarius general. As many know, he wrote the history of his own times, a work every way of uncommon value. By the best authorities, he has always been considered a most trustworthy and unexaggerating historian, except in some one or two particulars, not at all affecting the matter presently to be mentioned.

Now, in this history of his, Procopius mentions that, during the term of his prefecture at Constantinople, a great sea-monster was captured in the neighboring Propontis, or Sea of Marmora, after having destroyed vessels at intervals in those waters for a period of more than fifty years. A fact thus set down in substantial history cannot easily be gainsaid. Nor is there any reason it should be. Of what precise species this sea-monster was, is not mentioned. But as he destroyed ships, as well as for other reasons, he must have been a whale; and I am strongly inclined to think a sperm whale. And I will tell you why. For a long time I fancied that the sperm whale had been always unknown in the Mediterranean and the deep waters connecting with it. Even now I am certain that those seas are not, and perhaps never can be, in the present constitution of things, a place for his habitual gregarious resort. But further investigations have recently proved to me, that in modern times there have been isolated instances of the presence of the sperm whale in the Mediterranean. I am told, on good authority, that on the Barbary coast, a Commodore Davis of the British navy found the skeleton of a sperm whale. Now, as a vessel of war readily passes through the Dardanelles, hence a sperm whale could, by the same route, pass out of the Mediterranean into the Propontis.

In the Propontis, as far as I can learn, none of that peculiar substance called *brit* is to be found, the aliment of the right whale. But I have every reason to believe that the food of the sperm whale—squid or cuttle-fish—lurks at the bottom of that sea, because large creatures, but by no means the largest of that sort, have been found at its surface. If, then, you properly put these statements together, and reason upon them a bit, you will clearly perceive that, according to all human reasoning, Procopius's sea-monster, that for half a century stove the ships of a Roman Emperor, must in all probability have been a sperm whale.

Note

1. The following are extracts from Chace's narrative: "Every fact seemed to warrant me in concluding that it was anything but chance which directed his operations; he made two several attacks upon the ship, at a short interval between them, both of which, according to their direction, were calculated to do us the most injury, by being made ahead, and thereby combining the speed of the two objects for the shock; to effect which, the exact manoeuvres which he made were necessary. His aspect was most horrible, and such as indicated resentment and fury. He came directly from the shoal which we had just before entered, and in which we had struck three of his companions, as if fired with revenge for their sufferings." Again: "At all events, the whole circumstances taken together, all happening before my own eyes, and producing, at the time, impressions in my mind of decided, calculating mischief, on the part of the whale (many of which impressions I cannot now recall), induce me to be satisfied that I am correct in my opinion."

Here are his reflections some time after quitting the ship, during a black night in an open boat, when almost despairing of reaching any hospitable shore. "The dark ocean and swelling waters were nothing; the fears of being swallowed up by some dreadful tempest, or dashed upon hidden rocks, with all the other ordinary subjects of fearful contemplation, seemed scarcely entitled to a moment's thought; the dismal looking wreck, and *the horrid aspect and revenge of the whale*, wholly engrossed my reflections, until day again made its appearance."

In another place—p. 45,—he speaks of "*the mysterious and mortal attack of the animal.*"

Owen Chase, *Narrative of the Most Extraordinary and Distressing Shipwreck of the Whale-Ship* Essex

(New York: Gilley, 1821)

INTRODUCTION

Owen Chase (1797–1869) was the first mate of the whale-ship *Essex*; his book is the foundational work in the study of the *Essex* disaster and the whaling sources of *Moby-Dick*. Chase's account introduced the concept of a malignant whale deliberately seeking the destruction of an entire ship—a vessel displacing hundreds of tons and measuring about a 100 feet long (the whale himself was about 80 feet long and, by the estimate of those on board, could have weighed about 80 tons). Chase's book further provided Melville and his contemporaries with a detailed description of the sufferings leading to survival cannibalism. Frequently adapted, the account of the *Essex* became the archetypal whaling legend before *Moby-Dick*.

Melville read Chase's book and thought he had actually seen him while at sea on his first whale-ship, *Acushnet*, in the early 1840s, but his recollections are untrustworthy: "In all Melville's life," preeminent Melville scholar Hershel Parker has written in *Herman Melville: A Biography,* "he is not known to have so intricately tangled up so much misapprehension and misinformation as he did about Owen Chase." When Melville finally procured his own copy of Chase's book, he wrote copious notes in it, not only what he misremembered about Chase but also about Captain George Pollard Jr. (1791–1870). About the latter, Melville records, "Pollard himself returned to Nantucket, &

subsequently sailed on another whaling voyage to the Pacific, but he had not been in the Pacific long, when one night, his ship went ashore on unknown rocks, & was dashed to peaces [*sic*].” Melville’s notes are reported and analyzed in Parker and Hayford (2002), Parker (1996), the Northwestern-Newberry *Moby-Dick*, and Heffernan (1981).

Melville’s claim that he had seen “stupedly abbreviated” versions of the *Essex* story may be overly harsh. The only truly stupid statement about the incident is probably only an error of transcription, in Adam Hodgson’s *Letters from North America, Written during a Tour in the United States and Canada* (1824): “See a very interesting account of the wreck of a whale ship from Kentucky, in the Appendix D.” The appendix is hardly an abbreviation, taking up 26 pages. “The accuracy of the narrative,” Hodgson concluded, “I have reason to believe to be unquestionable. The father of one of the sufferers who died, was engaged to dine with me when the account reached Liverpool, and was prevented by the melancholy intelligence.”

Melville received his copy from Thomas Macy via Judge Lemuel Shaw in April 1851. Sealts #133; Bercaw #130. Modern edition: Heffernan (1981).

PRIMARY SOURCE

To the Reader

I am aware that the public mind has been already nearly sated with the private stories of individuals, many of whom had few, if any, claims to public attention, and the injuries which have resulted from the promulgation of fictitious histories, and in many instances, of journals entirely fabricated for the purpose, has had the effect to lessen the public interest in works of this description, and very much to undervalue the general cause of truth. It is, however, not the less important and necessary, that narratives should continue to be furnished that have their foundation in fact, and the subject of which embraces new and interesting matter in any department of the arts or sciences. When the motive is worthy, the subject and style interesting, affording instruction, exciting a proper sympathy, and withal disclosing new and astonishing traits of human character:—this kind of information becomes of great value to the philanthropist and philosopher, and is fully deserving of attention from every description of readers.

On the subject of the facts contained in this little volume, they are neither so extravagant, as to require the exercise of any great credulity to believe, nor, I trust, so unimportant or uninteresting, as to forbid an attentive perusal. It was my misfortune to be a considerable, if not a principal, sufferer, in the dreadful catastrophe that befell us; and in it, I not only lost all the little I had ventured, but my situation and the prospects of bettering it, that at one time seemed to smile upon me, were all in one short moment destroyed with it. The hope of obtaining something of remuneration, by giving a short history of my sufferings to the world, must therefore constitute my claim to public attention.

Preface

The increasing attention which is bestowed upon the whale fishery in the United States, has lately caused a very considerable commercial excitement; and no doubt it will become, if it be not at present, as important and general a branch of commerce as any belonging to our country. It is now principally confined to a very industrious and enterprising portion of the population of the States, many individuals of whom have amassed very rapid and considerable fortunes. It has enriched the inhabitants without bringing with it the usual corruptions and luxuries of a foreign trade; and those who are now most successful and conspicuous in it, are remarkable for the primitive simplicity, integrity, and hospitality of the island. This trade, if I may so call it, took its rise amongst the earliest settlers, and has gradually advanced to the extended, important, and lucrative state in which it now is, without any material interruption, and with very little competition until the present time. The late war temporally, but in a great degree affected its prosperity, by subjecting numerous fine vessels with their cargoes to capture and loss; but in its short continuance, it was not sufficient to divert the enterprise of the whalemen, nor to subdue the active energies of the capitalists embarked in it. At the conclusion of peace, those energies burst out afresh; and our sails now almost whiten the distant confines of the Pacific. The English have a few ships there; and the advantages which they possess over ours, it may be feared will materially affect our success, by producing in time a much more extensive and powerful competition. They are enabled to realize a greater profit from the demand and price of oil in their markets; and the encouragement afforded by parliament, not only in permitting the importation of it free of duty, but in granting a liberal bounty.

It is to be hoped that the wisdom of Congress will be extended to this subject; and that our present decided supremacy will not be lost for the want of a deserved government patronage.

Recent events have shown that we require a competent naval force in the Pacific, for the protection of this important and lucrative branch of commerce; for the want of which, many serious injuries and insults have been lately received, which have a tendency to retard its flourishing progress, and which have proved of serious consequence to the parties concerned.

During the late war, the exertions and intrepidity of Capt. Porter, were the means of saving a great deal of valuable property, which otherwise must have fallen into the hands of the enemy. His skilful, spirited, and patriotic conduct, on all occasions where he was called upon to act, imparted a protection and confidence to our countrymen, which completely fulfilled their expectations of him, and without doubt those of the government in sending him there.

Our ships usually occupy from two to three years in making a voyage. Occasionally, necessity obliges them to go into port for provisions, water, and repairs; in some cases, amongst mere savages, and in others, inhospitable people, from whom they are liable to every species of fraud, imposition, and force, which require some competent power to awe and redress. As long as the struggle between the patriots and royalists continues, or even should that speedily end—as long as young and instable governments, as there naturally must be for many years to come, exist there, our whalemen will continue to require that countenance and support which the importance and prosperity of the trade to them, and to the country, eminently entitle them. It is, undoubtedly, a most hazardous business, involving many incidental and unavoidable sacrifices, the severity of which it seems cruel to increase by the neglect or refusal of a proper protection.

The seamen employed in the fishery, and particularly those from Nantucket, are composed of the sons and connexions of the most respectable families on the island; and, unlike the majority of the class or profession to which they belong, they labour not only for their temporary subsistence, but they have an ambition and pride among them which seeks after distinguishment and promotion. Almost all of them enter the service with views of a future command; and submit cheerfully to the hardships and drudgery of the intermediate stations, until they become thoroughly acquainted with their business.

There are common sailors, boat-steerers, and harpooners: the last of these is the most honourable and important. It is in this station, that all the capacity of the young sailor is elicited; on the dexterous management of the harpoon, the line, and the lance, and in the adventurous positions which he takes alongside of his enemy, depends almost entirely the successful issue of his attack; and more real chivalry is not often exhibited on the deck of a battleship, than is displayed by these hardy sons of the ocean, in some of their gallant exploits among the whales. Nursed in the dangers of their business, and exposed to the continual hazards and hardships of all seasons, climates, and weathers, it will not be surprising if they should become a fearless set of people, and pre-eminent in all the requisites of good seamen. Two voyages are generally considered sufficient to qualify an active and intelligent young man for command; in which time, he learns from experience, and the examples which are set him, all that is necessary to be known.

While on this subject, I may be allowed to observe that it would not be an unprofitable task in a majority of our respectable shipmasters in the merchant service, to look into the principles of conduct, and study the economical [*sic*] management of the captains of our whale-ships. I am confident many serviceable hints could be gathered from the admirable system by which they regulate their concerns. They would learn, also, what respect is due to the character and standing of a captain of a whale-ship, which those of the merchant service affect so much to undervalue. If the post of danger be the post of honour; and if merit emanates from exemplary private character, uncommon intelligent, and professional gallantry, then is it due to a great majority of the shipmasters of Nantucket, that they should be held above the operations of an invidious and unjust distinction? It is a curious fact that one does exist; and it is equally an illiberal, as an undeserved reproach upon them, which time and an acquaintance with their merits must speedily wipe away.

Chapter I

The town of Nantucket, in the State of Massachusetts, contains about eight thousand inhabitants; nearly a third part of the population are quakers, and they are, taken together, a very industrious and enterprising people. On this island are owned about one hundred vessels, of all descriptions, engaged in the whale trade, giving constant employment and support to upwards of sixteen hundred hardy seaman, a class of people proverbial for their intrepidity. This fishery is not carried on to any extent from any other part of the United States, except from the town of New Bedford, directly opposite to Nantucket, where are owned probably twenty sail. A voyage generally lasts about two years and a half, and with an entire uncertainty of success. Sometimes they are repaid with speedy voyages and profitable cargoes, and at others they drag out a listless and disheartening cruise, without scarcely making the expenses of an outfit. The business is considered a very hazardous one, arising from unavoidable accidents, in carrying on an exterminating warfare against those great leviathans of the deep; and indeed a Nantucket man is on all occasions fully sensible of the honour and merit of his profession; no doubt because he knows that his laurels, like the soldier's, are plucked from the brink of danger. Numerous anecdotes are related of the whalemen of Nantucket; and stories of hair-breadth 'scapes, and sudden and wonderful preservation, are handed down amongst them, with the fidelity, and no doubt many of them with the characteristic fictions of the ancient legendary tales. A spirit of adventure amongst the sons of other relatives of these immediately concerned in it, takes possession of their minds at an early age; captivated with the tough stories of the elder seamen, and seduced, as well as the natural desire of seeing foreign countries, as by the hopes of gain, they launch forth six or eight thousand miles from home, into an almost untraversed ocean and spend from two to three years of their lives in scenes of constant peril, labour, and watchfulness. The profession is one of great ambition, and full of honourable excitement: a tame man is never known amongst them; and the coward is marked with that peculiar aversion, that distinguishes our public naval service. There are perhaps no people of superior corporeal powers; and it has been truly said of them, that they possess a natural aptitude, which seems rather the lineal spirit of their fathers, than the effects of any experience. The town itself, during the war, was (naturally to have been expected) on the decline; but with the return of peace it took a fresh start, and a spirit for carrying on the fishery received a renewed and very considerable excitement. Large capitals are now embarked; and some of the finest ships that our country can boast of are employed in it. The increased demand, within a few years past, from the spermaceti manufactories, has induced companies and individuals in different parts of the Union to become engaged in the business; and if the future consumption of the manufactured article bear any proportion to that of the few past years, this species of commerce will bid fair to become the most profitable and extensive that our country possesses. From the accounts of those who were in the early stages of the fishery concerned in it, it would appear, that the whales have been driven, like the beasts of the forest, before the march of civilization, into remote and more unfrequented seas, until now, they are followed by the enterprise and perseverance of our seamen, even to the distant coasts of Japan.

The ship Essex, commanded by captain George Pollard, junior, was fitted out at Nantucket, and sailed on the 12th day of August, 1819, for the Pacific Ocean, on a whaling voyage. Of this ship I was first mate. She had lately undergone a thorough repair in her upper works, and was at that time, in all respects, a sound, substantial vessel: she had a crew of twenty-one men, and was victualled and provided for two years and a half. We left the coast of America with a fine breeze, and steered for the Western Islands. On the second day out, while sailing moderately on our course in the Gulf Stream, a sudden squall of wind struck the ship from the SW, and knocked her completely on her beam-ends, stove one of our boats, entirely destroyed two others, and threw down the cambouse. We distinctly saw the approach of this gust, but miscalculated altogether as to the strength and violence of it. It struck the ship about three points off the weather quarter, at the moment that the man at the helm was in the act of putting her away to run before it. In an instant she was knocked down with her yards in the water; and before hardly a moment of time was allowed for reflection, she gradually came to the wind, and righted. The squall was accompanied with vivid flashes of lighting, and heavy and repeated claps of thunder. The whole ship's crew were, for a short time, thrown into the utmost consternation and confusion; but fortunately the violence of the squall was all contained in the first gust of the wind, and it soon gradually abated, and became fine weather again. We repaired our damage with little difficulty, and continued on our course, with the loss of the two boats. On the 30th of August we made the island of Floros, one of the western group called the Azores. We lay off and on the island for two days, during which time our boats landed and obtained a supply of vegetables and a few hogs: from this place we took the NE trade-wind, and in sixteen days made the Isle of May, one of the Cape de Verdes. As we were sailing along the shore of this island, we discovered a ship stranded on the beach, and from her appearance took her to be a whaler. Having lost two of our boats, and presuming that this vessel had probably some belonging to her that might have been saved, we determined to ascertain the name of the ship, and endeavor to supply if possible the loss of our boats from her. We accordingly stood in towards the port, or landing place. After a short time three men were discovered coming out to us in a whale boat. In a few moments they were alongside, and informed us that the wreck was the Archimedes of New York, captain George B. Coffin, which vessel had struck on a rock near the island about a fortnight previously; that all hands were saved by running the ship on shore, and that the captain and crew had gone home. We purchased the whale boat of these people, obtained some few more pigs, and again set sail. Our passage thence to Cape Horn was not distinguished for any incident worthy of note. We made the longitude of the Cape about the 18th of December, having experienced head winds for nearly the whole distance. We anticipated a moderate time in passing this noted land, from the season of the year at which we were there, being considered the most favourable; but instead of this, we experienced heavy westerly gales, and a most tremendous sea, that detained us off the Cape five weeks, before we had got sufficiently to the westward to enable us to put away. Of the passage of this famous Cape it may be observed, that strong westerly gales and a heavy sea are its almost universal attendants: the prevalence and constancy of this wind and sea necessarily produce a rapid current, by which vessels are set to leeward; and it is not without some favourable slant of wind that they can in many cases get round at all. The difficulties and dangers of the passage are proverbial; but as far as my own observation extends (and which the numerous reports of the whalemen corroborate), you can always rely upon a long and regular sea; and although the gales may be very strong and stubborn, as they undoubtedly are, they are not known to blow with the destructive violence that characterizes some of the tornadoes of the western Atlantic Ocean. On the 17th of January, 1820, we arrived at the island of St. Mary's lying on the coast of Chile, in latitude 36° 59' S. longitude 73° 41' W. This island is a sort of rendezvous for whalers, from which they obtain their wood and water, and between which and the main land (a distance of about ten miles) they frequently cruise for a species of whale called the right whale. Our object in going in there was merely to get the news. We sailed thence to the island of Massafuera, where we got some wood and fish, and thence for the cruising ground along the coast of Chile, in search of the spermaceti whale. We took there eight, which yielded us two hundred and fifty barrels of oil; and the season having by this time expired, we changed our cruising ground to the coast of Peru. We obtained there five hundred and fifty barrels. After going into the small port of Decamas, and replenishing our wood and water, on the 2nd of October we set sail for the Galapagos Islands. We came to anchor, and laid seven days off Hood's Island, one of the group; during which time we stopped a leak which we had discovered, and obtained three hundred turtle. We then visited Charles Island, where we procured sixty more. These turtle are a most delicious food, and average in weight generally about one hundred pounds, but many of them weigh upwards of eight hundred. With these, ships usually supply themselves for a great length of time and make a great saving of other provisions. They neither eat nor drink, nor is the least pains taken with them; they are strewed over the deck, thrown underfoot, or packed away in the hold, as it suits convenience. They will live upwards of a year without food or water, but soon die in a cold climate. We left Charles Island on the 23rd of October, and steered off to the westward, in search of whales. In latitude 1° 0' S. longitude 118° 0' W. on the 16th of November, in the afternoon, we lost a boat during our work in a shoal of whales. I was in the boat myself, with five others, and was standing in the fore part, with the harpoon in my hand, well braced, expecting every instant to catch sight of one of the shoal which we were in, that I might strike; but judge of my astonishment and dismay, at finding myself suddenly thrown up in the air, my companions scattered about me, and the boat fast filling with water. A whale had come up directly under her, and with one dash of his tail, had stove her bottom in, and strewed us in every direction around her. We, however, with little difficulty, got safely on the wreck, and clung there until one of the other boats which had been engaged in the shoal, came to our assistance, and took us off. Strange to tell, not a man was injured by this accident. Thus it happens very frequently in the whaling business, that boats are stove; oars, harpoons, and lines broken; ankles and wrists sprained; boats upset, and whole crews left for hours in the water, without any of these accidents extending to the loss of life. We are so much accustomed to the continual recurrence of such scenes as these, that we become familiarized to them, and consequently always feel that confidence and self-possession, which teaches us every expedient in danger, and inures the body, as well as the mind, to fatigue, privation, and peril, in frequent cases exceeding belief. It is this danger and hardship that makes the sailor; indeed it is the distinguishing qualification amongst us; and it is a common boast of the whaleman, that he has escaped from sudden and apparently inevitable destruction oftener than his fellow. He is accordingly valued on this account, without much reference to other qualities.

Chapter II

I have not been able to recur to the scenes which are now to become the subject of description, although a considerable time has elapsed without feeling a mingled emotion of horror and astonishment at the almost incredible destiny that has preserved me and my surviving companions from a terrible death. Frequently, in my reflections on the subject, even after this lapse of time, I find myself shedding tears of gratitude for our deliverance, and blessing God, by whose divine aid and protection we were conducted through a series of unparalleled suffering and distress, and restored to the bosoms of our families and friends. There is no knowing what a stretch of pain and misery the human mind is capable of contemplating, when it is wrought upon by the anxieties of preservation; nor what pangs and weaknesses the body is able to endure, until they are visited upon it; and when at last deliverance comes, when the dream of hope is realized, unspeakable gratitude takes possession of the soul, and tears of joy choke the utterance. We require to be taught in the school of some signal suffering, privation, and despair, the great lessons of constant dependence upon an almighty forbearance and mercy. In the midst of the wide ocean, at night, when the sight of the heavens was shut out, and the dark tempest came upon us; then it was, that we felt ourselves ready to exclaim, "Heaven have mercy upon us, for nought but that can save us now." But I proceed to the recital.—On the 20th of November (cruising in latitude 0° 40' S. longitude 119° 0' W.), a shoal of whales was discovered off the lee-bow. The weather at this time was extremely fine and clear, and it was about 8 o'clock in the morning, that the man at the mast-head gave the usual cry of, "there she blows." The ship was immediately put away, and we ran down in the direction for them. When we had got within half a mile of the place where they were observed, all our boats were lowered down, manned, and we started in pursuit of them. The ship, in the mean time, was brought to the wind, and the main-top-sail hove aback, to wait for us. I had the harpoon in the second boat; the captain preceded me in the first. When I arrived at the spot where we calculated they were, nothing was at first to be seen. We lay on our oars in anxious expectation of discovering them come up somewhere near us. Presently one rose, and spouted a short distance ahead of my boat; I made all speed towards it, came up with, and struck it; feeling the harpoon in him, he threw himself, in an agony, over towards the boat (which at that time was up alongside of him), and giving a severe blow with his tail, struck the boat near the edge of the water, amidships, and stove a hole in her. I immediately took up the boat hatchet, and cut the line, to disengage the boat from the whale, which by this time was running off with great velocity. I succeeded in getting clear of him, with the loss of the harpoon and line; and finding the water to pour fast in the boat. I hastily stuffed three or four of our jackets in the hole, ordered one man to keep constantly bailing, and the rest to pull immediately for the ship; we succeeded in keeping the boat free, and shortly gained the ship. The captain and the second mate, in the other two boats, kept up the pursuit, and soon struck another whale. They being at this time a considerable distance to leeward, I went forward, braced around the main-yard, and put the ship off in a direction for them; the boat which had been stove was immediately hoisted in, and after examining the hole, I found that I could, by nailing a piece of canvass over it, get her ready to join in a fresh pursuit, sooner than by lowering down the other remaining boat which belonged to the ship. I accordingly turned her over upon the quarter, and was in the act of nailing on the canvass, when I observed a very large spermaceti whale, as well as I could judge, about eighty-five feet in length; he broke water about twenty rods off our weather-bow, and was lying quietly, with his head in a direction for the ship. He spouted two or three times, and then disappeared. In less than two or three seconds he came up again, about the length of the ship off, and made directly for us, at the rate of about three knots. The ship was then going with about the same velocity. His appearance and attitude gave us at first no alarm; but while I stood watching his movements, and observing him but a ship's length off, coming down for us with great celerity, I involuntarily ordered the boy at the helm to put it hard up; intending to sheer off and avoid him. The words were scarcely out of my mouth, before he came down upon us with full speed, and struck the ship with his head, just forward of the fore-chains; he gave us such an appalling and tremendous jar, as nearly threw us all on our faces. The ship brought up as suddenly and violently as if she had struck a rock and trembled for a few seconds like a leaf.

We looked at each other with perfect amazement, deprived almost of the power of speech. Many minutes elapsed before we were able to realize the dreadful accident; during which time he passed under the ship, grazing her keel as he went along, came up underside of her to leeward, and lay on the top of the water (apparently stunned with the violence of the blow), for the space of a minute; he then suddenly started off, in a direction to leeward. After a few moments' reflection, and recovering, in some measure, from the sudden consternation that had seized us, I of course concluded that he had stove a hole in the ship, and that it would be necessary to set the pumps going. Accordingly they were rigged, but had not been in operation more than one minute, before I perceived the head of the ship to be gradually settling down in the water; I then ordered the signal to be set for the other boats, which scarcely had I dispatched, before I again discovered the whale, apparently in convulsions, on the top of the water, about one hundred rods to leeward. He was enveloped in the foam of the sea, that his continual and violent thrashing about in the water had created around him, and I could distinctly see him smite his jaws together, as if distracted with rage and fury. He remained a short time in this situation, and then started off with great velocity, across the bows of the ship, to windward. By this time the ship had settled down a considerable distance in the water, and I gave her up as lost. I however, ordered the pumps to be kept constantly going, and endeavoured to collect my thoughts for the occasion. I turned to the boats, two of which we then had with the ship, with an intention of clearing them away, and getting all things ready to embark in them, if there should be no other resource left; and while my attention was thus engaged for a moment, I was aroused with the cry of a man at the hatchway, "here he is—he is making for us again." I turned around, and saw him about one hundred rods directly ahead of us, coming down apparently with twice his ordinary speed, and to me at that moment, it appeared with tenfold fury and vengeance in his aspect. The surf flew in all directions about him, and his course towards us was marked by a white foam of a rod in width, which he made with the continual violent thrashing of his tail; his head was about half out of water, and in that way he came upon, and again struck the ship. I was in hopes when I descried him making for us, that by a dexterous movement of putting the ship away immediately, I should be able to cross the line of his approach, before he could get up to us, and thus avoid, what I knew, if he should strike us again, would prove our inevitable destruction. I bawled out to the helmsman, "hard up!" but she had not fallen off more than a point, before we took the second shock. I should judge the speed of the ship to have been at this time about three knots, and that of the whale about six. He struck her to windward, directly under the cathead, and completely stove in her bows. He passed under the ship again, went off to leeward, and we saw no more of him. Our situation at this juncture can be more readily imagined than described. The shock to our feelings was such, as I am sure none can have an adequate conception of, that were not there: the misfortune befell us at a moment when we least dreamt of any accident; and from the pleasing anticipations we had formed, of realizing the certain profits of our labour, we were dejected by a sudden, most mysterious, and overwhelming calamity. Not a moment, however, was to be lost in endeavouring to provide for the extremity to which it was now certain we were reduced. We were more than a thousand miles from the nearest land, and with nothing but a light open boat, as the resource of safety for myself and companions. I ordered the men to cease pumping, and every one to provide for himself, seizing a hatchet at the same time, I cut away the lashings of the spare boat, which lay bottom up, across two spars directly over the quarter deck, and cried out to those near me, to take her as she came down. They did so accordingly, and bore her on their shoulders as far as the waist of the ship. The steward had in the mean time gone down into the cabin twice, and saved two quadrants, two practical navigators, and the captain's trunk and mine; all which were hastily thrown into the boat, as she lay on the deck, with the two compasses which I snatched from the binnacle. He attempted to descend again; but the water by this time had rushed in, and he returned without being able to effect his purpose. By the time we had got the boat to the waist, the ship had filled with water, and was going down on her beam-ends: we shoved our boat as quickly as possible from the plank-shear into the water, all hands jumping in her at the same time, and launched off clear of the ship. We were scarcely two boat's lengths distant from her, when she fell over to windward, and settled down in the water.

Amazement and despair now wholly took possession of us. We contemplated the frightful situation the ship lay in, and thought with horror upon the sudden and dreadful calamity that had overtaken us. We looked upon each other, as if to gather some consolatory sensation from an interchange of sentiments, but every countenance was marked with the paleness of despair. Not a word was spoken for several minutes by any of us; all appeared to be bound in a spell of stupid consternation; and from the time we were first attacked by the whale, to the period of the fall of the ship, and of our leaving her in the boat, more than ten minutes could not certainly have elapsed! God only knows in what way, or by what means, we were enabled to accomplish in that short time what we did, the cutting away and transporting the boat from where she was deposited would of itself, in ordinary circumstances have consumed as much time as that, if the whole ship's crew had been employed in it. My companions had not saved a single article but what they had on their backs; but to me it was a source of infinite satisfaction, if any such could be gathered from the horrors of our gloomy situation, that we had been fortunate enough to have preserved our compasses, navigators, and quadrants. After the first shock of my feelings was over, I enthusiastically contemplated them as the probable instruments of our salvation; without them all would have been dark and hopeless. Gracious God! What a picture of distress and suffering now presented itself to my imagination. The crew of the ship were saved, consisting of twenty human souls. All that remained to conduct these twenty beings through the stormy terrors of the ocean, perhaps many thousand miles, were three open light boats. The prospect of obtaining any provisions or water from the ship, to subsist upon during the time, was at least now doubtful. How many long and watchful nights, thought I, are to be passed? How many tedious days of partial starvation are to be endured, before the least relief of mitigation of our sufferings can be reasonably anticipated? We lay at this time in our boat, about two ship's lengths off from the wreck, in perfect silence, calmly contemplating her situation, and absorbed in our own melancholy reflections, when the other boats were discovered rowing up to us. They had but shortly before discovered that some accident had befallen us, but of the nature of which they were entirely ignorant. The sudden and mysterious disappearance of the ship was first discovered by the boat-steerer in the captain's boat, and with a horror-struck countenance and voice, he suddenly exclaimed, "Oh, my God! where is the Ship?" Their operations upon this were instantly suspended, and a general cry of horror and despair burst from the lips of every man, as their looks were directed for her, in vain, over every part of the ocean. They immediately made all haste towards us. The captain's boat was the first that reached us. He stopped about a boat's length off, but had no power to utter a single syllable: he was so completely overpowered with the spectacle before him, that he sat down in his boat, pale and speechless. I could scarcely recognize his countenance, he appeared to be so much altered, awed, and overcome, with the oppression of his feelings, and the dreadful reality that lay before him. He was in a short time however enabled to address the inquiry to me, "My God, Mr. Chase, what is the matter?" I answered, "We have been stove by a whale." I then briefly told him the story. After a few moment's reflection he observed, that we must cut away her masts, and endeavour to get something out of her to eat. Our thoughts were now all accordingly bent on endeavours to save from the wreck whatever we might possibly want, and for this purpose we rowed up and got on to her. Search was made for every means of gaining access to her hold; and for this purpose the lanyards were cut loose, and with our hatchets we commenced to cut away the masts, that she might right up again, and enable us to scuttle her decks. In doing which we were occupied about three quarters of an hour owing to our having no axes, nor indeed any other instruments, but the small hatchets belonging to the boats. After her masts were gone she came up about two-thirds of the way upon an even keel. While we were employed about the masts the captain took his quadrant, shoved off the ship, and got an observation. We found ourselves in latitude 0° 40' S. longitude 119° W. We now commenced to cut a hole through the planks, directly above two large casks of bread, which most fortunately were between decks, in the waist of the ship, and which being in the upper side, when she upset, we had strong hopes was not wet. It turned out according to our wishes, and from these casks we obtained six hundred pounds of hard bread. Other parts of the deck were then scuttled, and we got without difficulty as much fresh water as we dared to take in the boats, so that each was supplied with about sixty-five gallons; we got also from one of the lockers a musket, a small canister of powder, a couple of files, two rasps, about two pounds of boat nails, and a few turtle. In the afternoon the wind came on to blow a strong breeze; and having obtained every thing that occurred to us could then be got out, we began to make arrangements for our safety during the night. A boat's line was made fast to the ship, and to the other end of it one of the boats was moored; at about fifty fathoms to leeward; another boat was then attached to the first one, about eight fathoms astern; and the third boat, the like distance astern of her. Night came on just as we had finished our operations; and such a night as it was to us! So full of feverish and distracting inquietude, that we were deprived entirely of rest. The wreck was constantly before my eyes. I could not, by any effort, chase away the horrors of the preceding day from my mind: they haunted me the live-long night. My companions—some of them were like sick women; they had no idea of the extent of their deplorable situation. One or two slept unconcernedly, while others wasted the night in unavailing murmurs. I now had full leisure to examine, with some degree of coolness, the dreadful circumstances of our disaster. The scenes of yesterday passed in such quick succession in my mind that it was not until after many hours of severe reflection that I was able to discard the idea of the catastrophe as a dream. Alas! It was one from which there was no awaking; it was too certainly true, that but yesterday we had existed as it were, and in one short moment had been cut off from the hopes and prospects of the living! I have no language to paint out the horrors of our situation. To shed tears was indeed altogether unavailing, and withal unmanly; yet I was not able to deny myself the relief they served to afford me. After several hours of idle sorrow and repining I began to reflect upon the accident, and endeavoured to realize by what unaccountable destiny or design (which I could not at first determine), this sudden and most deadly attack had been made upon us: by an animal, too, never before suspected of premeditated violence, and proverbial for its insensibility and inoffensiveness. Every fact seemed to warrant me in concluding that it was any thing but chance which directed his operations; he made two several attacks upon the ship, at a short interval between them, both of which, according to their direction, were calculated to do us the most injury, by being made ahead, and thereby combining the speed of the two objects for the shock; to effect which, the exact manoeuvres which he made were necessary. His aspect was most horrible, and such as indicated resentment and fury. He came directly from the shoal which we had just before entered, and in which we had struck three of his companions, as if fired with revenge for their sufferings. But to this it may be observed, that the mode of fighting which they always adopt is either with repeated strokes of their tails, or snapping of their jaws together; and that a case, precisely similar to this one, has never been heard of amongst the oldest and most experienced whalers. To this I would answer, that the structure and strength of the whale's head is admirably designed for this mode of attack; the most prominent part of which is almost as hard and as tough as iron; indeed, I can compare it to nothing else but the inside of a horse's hoof, upon which a lance or harpoon would not make the slightest impression. The eyes and ears are removed nearly one-third the length of the whole fish, from the front part of the head, and are not in the least degree endangered in this mode of attack. At all events, the whole circumstances taken together, all happening before my own eyes, and producing, at the time, impressions in my mind of decided, calculated mischief, on the part of the whale (many of which impressions I cannot now recall), induce me to be satisfied that I am correct in my opinion. It is certainly, in all its bearings, a hitherto unheard of circumstance, and constitutes, perhaps, the most extraordinary one in the annals of the fishery.

Chapter III

November 21st. The morning dawned upon our wretched company. The weather was fine, but the wind blew a strong breeze from the SE, and the sea was very rugged. Watches had been kept up during the night, in our respective boats, to see that none of the spars or other articles (which continued to float out of the wreck) should be thrown by the surf against, and injure the boats. At sunrise, we began to think of doing something; what, we did not know; we cast loose our boats, and visited the wreck, to see if any thing more of consequence could be preserved, but every thing looked cheerless and desolate, and we made a long and vain search for any useful article; nothing could be found but a few turtle; of these we had enough already; or at least, as many as could be safely stowed in the boats, and we wandered around in every part of the ship in a sort of vacant idleness for the greater part of the morning. We were presently aroused to a perfect sense of our destitute and forlorn condition, by thoughts of the means which we had for our subsistence, the necessity of not wasting our time, and of endeavouring to seek some relief wherever God might direct us. Our thoughts, indeed, hung about the ship, wrecked and sunken as she was, and we could scarcely discard from our minds the idea of her continuing protection. Some great efforts in our situation were necessary, and a great deal of calculation important, as it concerned the means by which our existence was to be supported during, perhaps, a very long period, and a provision for our eventual deliverance. Accordingly, by agreement, all set to work in stripping off the light sails of the ship, for sails to our boats; and the day was consumed in making them up and fitting them. We furnished ourselves with masts and other light spars that were necessary, from the wreck. Each boat was rigged with two masts, to carry a flying-jib and two spirit-sails; the spirit-sails were made so that two reefs could be taken in them, in case of heavy blows. We continued to watch the wreck for any serviceable articles that might float from her, and kept one man during the day, on the stump of her foremast, on the look out for vessels. Our work was very much impeded by the increase of the wind and sea, and the surf breaking almost continually into the boats, gave us many fears that we should not be able to prevent our provisions from getting wet; and above all served to increase the constant apprehensions that we had, of the insufficiency of the boats themselves, during the rough weather that we should necessarily experience. In order to provide as much as possible against this, and withal to strengthen the slight materials of which the boats were constructed, we procured from the wreck some light cedar boards (intended to repair boats in case of accidents), with which we built up additional sides, about six inches above the gunwale; these, we afterwards found, were of infinite service for the purpose for which they were intended; in truth, I am satisfied we could never have been preserved without them; the boats must otherwise have taken in so much water that all the efforts of twenty such weak, starving men as we afterwards came to be, would not have sufficed to keep her free; but what appeared most immediately to concern us, and to command all our anxieties, was the security of our provisions from the salt water. We disposed of them under a covering of wood, that whale-boats have at either end of them, wrapping it up in several thicknesses of canvass. I got an observation to-day, by which I found we were in latitude 0° 61' S. longitude 119° 30' W. having been driven by the winds a distance of forty-nine miles the last twenty-four hours; by this it would appear that there must have been a strong current, setting us to the NW during the whole time. We were not able to finish our sails in one day; and many little things preparatory to taking a final leave of the ship were necessary to be attended to, but evening came and put an end to our labours. We made the same arrangements for mooring the boats in safety, and consigned ourselves to the horrors of another tempestuous night. The wind continued to blow hard, keeping up a heavy sea, and veering around from SE to E, and E.SE. As the gloom of night approached, and obliged us to desist from that employment, which cheated us out of some of the realities of our situation, we all of us again became mute and desponding: a considerable degree of alacrity had been manifested by many the preceding day, as their attention had been wholly engaged in scrutinizing the wreck, and in constructing the sails and spars for the boats; but when they ceased to be occupied, they passed to a sudden fit of melancholy, and the miseries of their situation came upon them with such force, as to produce spells of extreme debility, approaching almost to fainting. Our provisions were scarcely touched—the appetite was entirely gone; but as we had a great abundance of water, we indulged in frequent and copious draughts, which our parched mouths seemed continually to need. None asked for bread. Our continued state of anxiety during the night, excluded all hopes of sleep; still (although the solemn fact had been before me for nearly two days), my mind manifested the utmost repugnance to be reconciled to it; I laid down in the bottom of the boat, and resigned myself to reflection; my silent prayers were offered up to the God of mercy, for that protection which we stood so much in need of. Sometimes, indeed, a light hope would dawn, but then, to feel such an utter dependence on and consignment to chance alone for aid and rescue, would chase it again from my mind. The wreck—the mysterious and mortal attack of the animal—the sudden prostration and sinking of the vessel—our escape from her, and our then forlorn and almost hapless destiny, all passed in quick and perplexing review in my imagination; wearied with the exertion of the body and mind, I caught, near morning, an hour's respite from my troubles, in sleep.

November 22nd. The wind remained the same, and the weather continued remarkably fine. At sunrise, we again hauled our boats up, and continued our search for articles that might float out. About 7 o'clock, the deck of the wreck began to give way, and every appearance indicated her speedy dissolution; the oil had bilged in the hold, and kept the surface of the sea all around us completely covered with it; the bulk-heads were all washed down, and she worked in every part of her joints and seams, with the violent and continual breaking of the surf over her. Seeing, at last, that little or nothing further could be done by remaining with the wreck, and as it was all important that while our provisions lasted, we should make the best possible use of time, I rowed up to the captain's boat, and asked him what he intended to do. I informed him that the ship's decks had bursted up, and that in all probability she would soon go to pieces; that no further purpose could be answered, by remaining longer with her, since nothing more could be obtained from her; and that it was my opinion, no time should be lost in making the best of our way towards the nearest land. The captain observed, that he would go once more to the wreck, and survey her, and after waiting until 12 o'clock for the purpose of getting an observation, would immediately after determine. In the mean time, before noon all our sails were completed, and the boats otherwise got in readiness for our departure. Our observation now proved us to be in latitude 0° 13' N. longitude 120° 00' W. as near as we could determine it, having crossed the equator during the night, and drifted nineteen miles. The wind had veered considerably to the eastward, during the last twenty-four hours. Our nautical calculations having been completed, the captain, after visiting the wreck, called a council, consisting of himself and the first and second mates, who all repaired to his boat, to interchange opinions, and devise the best means for our security and preservation. There were, in all of us, twenty men; six of whom were blacks, and we had three boats. We examined our navigators, to ascertain the nearest land, and found it was the Marquesas Islands. The Society Islands were next; these islands we were entirely ignorant of; if inhabited, we presumed they were by savages, from whom we had as much to fear, as from the elements, or even death itself. We had no charts from which our calculations might be aided, and were consequently obliged to govern ourselves by the navigators alone; it was also the captain's opinion, that this was the season of the hurricanes which prevail in the vicinity of the Sandwich Islands, and that consequently it would be unsafe to steer for them. The issue of our deliberations was, that, taking all things into consideration, it would be most advisable to shape our course by the wind, to the southward, as far as 25°; or 26°; S. latitude, fall in with the variable winds, and then, endeavour to get eastward to the coast of Chile or Peru. Accordingly, preparations were made for our immediate departure; the boat which it was my fortune, or rather misfortune to have, was the worst of the three; she was old and patched up, having been stove a number of times, during the cruise. At best, a whale-boat is an extremely frail thing; the most so of any other kind of boat; they are what is called clinker built, and constructed of the lightest materials, for the purpose of being rowed with the greatest possible celerity, according to the necessities of the business for which they are intended. Of all species of vessels, they are the weakest, and most fragile, and possess but one advantage over any other—that of lightness and buoyancy, that enables them to keep above the dash of the sea, with more facility than heavier ones. This qualification is, however, preferable to that of any other, and, situated as we then were, I would not have exchanged her, old and crazy as she was, for even a ship's launch. I am quite confident, that to this quality of our boats we most especially owed our preservation, through the many days and nights of heavy weather, that we afterwards encountered. In consideration of my having the weakest boat; six men were allotted to it; while those of the captain and second mate, took seven each; and at half past 12 we left the wreck, steering our course, with nearly all sail set, S.SE. At four o'clock, in the afternoon we lost sight of her entirely. Many were the lingering and sorrowful looks we cast behind us.

It has appeared to me often since to have been, in the abstract, an extreme weakness and folly, on our parts, to have looked upon our shattered and sunken vessel with such an excessive fondness and regret; but it seemed as if in abandoning her we had parted with all hope, and were bending our course away from her, rather by some dictate of despair. We agreed to keep together, in our boats, as nearly as possible to afford assistance in cases of accident, and to render our reflections less melancholy by each other's presence. I found it on this occasion true, that misery does indeed love company; unaided, and unencouraged by each other, there were with us many whose weak minds, I am confident, would have sunk under the dismal retrospections of the past catastrophe, and who did not possess either sense or firmness enough to contemplate our approaching destiny, without the cheering of some more determined countenance than their own. The wind was strong all day; and the sea ran very high, our boat taking in water from her leaks continually, so that we were obliged to keep one man constantly bailing. During the night the weather became extremely rugged, and the sea every now and then broke over us. By agreement, we were divided into two watches; one of which was to be constantly awake, and doing the labours of the boat, such as bailing; setting, taking in, and trimming the sails. We kept our course very well together during this night, and had many opportunities of conversation with the men in the other boats, wherein the means and prospects of our deliverance were variously considered; it appeared from the opinions of all, that we had most to hope for in the meeting with some vessel, and most probably some whale ship, the great majority of whom, in those seas, we imagined were cruising about the latitude we were then steering for; but this was only a hope, the realization of which did not in any degree depend on our own exertions, but on chance alone. It was not, therefore, considered prudent, by going out of our course, with the prospect of meeting them, to lose sight, for one moment, of the strong probabilities which, under Divine providence, there were of our reaching land by the route we had prescribed to ourselves; as that depended, most especially, on a reasonable calculation, and on our own labours, we conceived that our provision and water, on a small allowance, would last us sixty days; that with the trade-wind, on the course we were then lying, we should be able to average the distance of a degree a day, which, in 26 days, would enable us to attain the region of the variable winds, and then, in thirty more, at the very utmost, should there be any favour in the elements, we might reach the coast. With these considerations we commenced our voyage; the total failure of all which, and the subsequent dismal distress and suffering, by which we were overtaken, will be shown in the sequel. Our allowance of provision at first consisted of bread; one biscuit, weighing about one pound three ounces, and half a pint of water a day, for each man. This small quantity (less than one third which is required by an ordinary person), small as it was, we however took without murmuring, and, on many an occasion afterwards, blest God that even this pittance was allowed to us in our misery. The darkness of another night overtook us, and after having for the first time partook of our allowance of bread and water, we laid our weary bodies down in the boat, and endeavoured to get some repose. Nature became at last worn out with the watchings and anxieties of the two preceding nights, and sleep came insensibly upon us. No dreams could break the strong fastenings of forgetfulness in which the mind was then locked up; but for my own part, my thoughts so haunted me that this luxury was yet a stranger to my eyes; every recollection was still fresh before me, and I enjoyed but a few short and unsatisfactory slumbers, caught in the intervals between my hopes and my fears. The dark ocean and swelling waters were nothing; the fears of being swallowed up by some dreadful tempest, or dashed upon hidden rocks, with all the other ordinary subjects of fearful contemplation, seemed scarcely entitled to a moment's thought; the dismal looking wreck, and the horrid aspect and revenge of the whale, wholly engrossed my reflections, until day again made its appearance.

November 23rd. In my chest, which I was fortunate enough to preserve, I had several small articles, which we found of great service to us; among the rest, some eight or ten sheets of writing paper, a lead pencil, a suit of clothes, three small fish hooks, a jack-knife, besides other useful purposes, served us as a razor. It was with much difficulty, however, that I could keep any sort of record, owing to the incessant rocking and unsteadiness of the boat, and the continual dashing of the spray of the sea over us. The boat contained, in addition to the articles enumerated, a lantern, tinderbox, and two or three candles, which belonged to her, and with which they are kept always supplied, while engaged in taking whale. In addition to all which, the captain had saved a musket, two pistols, and a canister, containing about two pounds of gunpowder; the latter he distributed in equal proportions between the three boats, and gave the second mate and myself each a pistol. When morning came we found ourselves quite near together, and the wind had considerably increased since the day before; we were consequently obliged to reef our sales and although we did not apprehend any very great danger from the then violence of the wind, yet it grew to be very uncomfortable in the boats, from the repeated dashing of the waves, that kept our bodies constantly wet with the salt spray. We, however, stood along our course until twelve o'clock, when we got an observation, as well as we were able to obtain one, while the water flew all over us, and the sea kept the boat extremely unsteady. We found ourselves this day in latitude 0° 58' S. having repassed the equator. We abandoned the idea altogether of keeping any correct longitudinal reckoning, having no glass, nor log-line. The wind moderated in the course of the afternoon a little, but at night came on to blow again almost a gale. We began now to tremble for our little barque; she was so ill calculated, in point of strength, to withstand the racking of the sea, while it required the constant labours of one man to keep her free of water. We were surrounded in the afternoon with porpoises that kept playing about us in great numbers, and continued to follow us during the night.

November 24th. The wind had not abated any since the preceding day, and the sea had risen to be very large, and increased, if possible, the extreme uncomfortableness of our situation. What added more than any thing else to our misfortunes, was, that all our efforts for the preservation of our provisions proved, in a great measure, ineffectual; a heavy sea broke suddenly into the boat, and before we could snatch it up, damaged some part of it; by timely attention, however, and great caution, we managed to make it eatable, and to preserve the rest from a similar casualty. This was a subject of extreme anxiety to us; the expectation, poor enough of itself indeed, upon which our final rescue was founded, must change at once to utter hopelessness, deprived of our provisions, the only means of continuing us in the exercise, not only of our manual powers, but in those of reason itself; hence, above all other things, this was the object of our utmost solicitude and pains.

We ascertained, the next day, that some of the provisions in the captain's boat had shared a similar fate during the night; both which accidents served to arouse us to a still stronger sense of our slender reliance upon the human means at our command, and to show us our utter dependence on that divine aid which we so much the more stood in need of.

November 25th. No change of wind had yet taken place, and we experienced the last night the same wet and disagreeable weather of the preceding one. About eight o'clock in the morning we discovered that the water began to come fast in our boat, and in a few minutes the quantity increased to such a degree as to alarm us considerably for our safety; we commenced immediately a strict search in every part of her to discover the leak, and, after tearing up the ceiling or floor of the boat near the bows, we found it proceeded from one of the streaks or outside boards having bursted off there; no time was to be lost in devising some means to repair it. The great difficulty consisted in its being in the bottom of the boat, and about six inches from the surface of the water; it was necessary, therefore, to have access to the outside, to enable us to fasten it on again: the leak being to leeward, we hove about, and lay to on the other tack which brought it then nearly out of water; the captain, who was at the time ahead of us, seeing us manoeuvring to get the boat about, shortened sail, and presently tacked, and ran down to us. I informed him of our situation, and he came immediately alongside to our assistance. After directing all the men in the boat to get on one side, the other, by that means, heeled out of the water a considerable distance, and, with a little difficulty, we then managed to drive in a few nails, and secured it, much beyond our expectations. Fears of no ordinary kind were excited by this seemingly small accident. When it is recollected to what a slight vessel we had committed ourselves; our means of safety alone consisting in her capacity of endurance for many weeks, in all probability, yet to come, it will not be considered strange that this little accident should not only have damped our spirits considerably, but have thrown a great gloominess over the natural prospects of our deliverance. On this occasion, too, were we enabled to rescue ourselves from inevitable destruction by the possession of a few nails, without which (had it not been our fortune to save some from the wreck), we would, in all human calculation, have been lost: we were still liable to a recurrence of the same accident, perhaps to a still worse one, as, in the heavy and repeated racking of the swell, the progress of our voyage would serve but to increase the incapacity and weakness of our boat, and the starting of a single nail in her bottom would most assuredly prove our certain destruction. We wanted not this additional reflection, to add to the miseries of our situation.

November 26th. Our suffering, heaven knows, were now sufficiently increased, and we looked forward, not without an extreme dread, and anxiety, to the gloomy and disheartening prospect before us. We experienced a little abatement of wind and rough weather today, and took the opportunity of drying the bread that had been wet the day previously; to our great joy and satisfaction also, the wind hauled out to E.NE. and enabled us to hold a much more favourable course; with these exceptions, no circumstance of any considerable interest occurred in the course of this day.

The 27th of November was alike undistinguished for any incident worthy of note; except that the wind again veered back to E. and destroyed the fine prospect we had entertained, of making a good run for several days to come.

November 28th. The wind hauled still further to the southward, and obliged us to fall off our course to S. and commenced to blow with such violence, as to put us again under short sail; the night set in extremely dark, and tempestuous, and we began to entertain fears that we should be separated. We however, with great pains, managed to keep about a ship's length apart, so that the white sails of our boats could be distinctly discernable. The captain's boat was but a short distance astern of mine, and that of the second mate a few rods to leeward of his. At about 11 o'clock at night, having laid down to sleep, in the bottom of the boat. I was suddenly awakened by one of my companions, who cried out, that the captain was in distress, and was calling on us for assistance. I immediately aroused myself, and listened a moment, to hear if any thing further should be said, when the captain's loud voice arrested my attention. He was calling to the second mate, whose boat was nearer to him than mine. I made all haste to put about, ran down to him, and inquired what was the matter; he replied, "I have been attacked by an unknown fish, and he has stove my boat." It appeared, that some large fish had accompanied the boat for a short distance, and had suddenly made an unprovoked attack upon her, as nearly as they could determine, with his jaws; the extreme darkness of the night prevented them from distinguishing what kind of animal it was, but they judged it to be about twelve feet in length, and one of the killer-fish species. After having struck the boat once, he continued to play about her, on every side, as if manifesting a disposition to renew the attack, and did a second time strike the bows of the boat, and split her stem. They had no other instrument of offence, but the spirit-pole (a long slender piece of wood, by which the peak of the sail is extended), with which, after repeated attempts to destroy the boat, they succeeded in beating him off. I arrived, just as he had discontinued his operations, and disappeared. He had made a considerable breach in the bows of the boat, through which the water had began [*sic*] to pour fast; and the captain, imagining matters to be considerably worse than they were, immediately took measures to remove his provisions into the second mate's boat and mine, in order to lighten his own, and by that means, and constant bailing, to keep her above water until daylight should enable him to discover the extent of the damage, and to repair it. The night was spissy darkness itself; the sky was completely overcast, and it seemed to us as if fate was wholly relentless, in pursuing us with such a cruel complication of disasters. We were not without our fears that the fish might renew his attack, some time during the night, upon one of the other boats, and unexpectedly destroy us; but they proved entirely groundless, as he was never afterwards seen. When daylight came, the wind again favoured us a little, and we all lay to, to repair the broken boat; which was effected by nailing on thin strips of boards in the inside; and having replaced the provisions, we proceeded again on our course. Our allowance of water, which in the commencement, merely served to administer to the positive demands of nature, became now to be insufficient; and we began to experience violent thirst, from the consumption of the provisions that had been wet with the salt water, and dried in the sun; of these we were obliged to eat first, to prevent their spoiling; and we could not, nay, we did not dare, to make any encroachments on our stock of water. Our determination was, to suffer as long as human patience and endurance would hold out, having only in view, the relief that would be afforded us, when the quantity of wet provisions should be exhausted. Our extreme sufferings here first commenced. The privation of water is justly ranked among the most dreadful of the miseries of our life; the violence of raving thirst has no parallel in the catalogue of human calamities. It was our hard lot, to have felt this in its extremest force, when necessity subsequently compelled us to seek resource from one of the offices of nature. We were not, at first, aware of the consequences of eating this bread; and it was not until the fatal effects of it had shown themselves to a degree of oppression, that we could divine the cause of our extreme thirst. But, alas! There was no relief. Ignorant, or instructed of the fact, it was alike immaterial; it composed a part of our subsistence, and reason imposed upon us the necessity of its immediate consumption, as otherwise it would have been lost to us entirely.

November 29th. Our boats appeared to be growing daily more frail and insufficient; the continual flowing of the water into them, seemed increased, without our being able to assign it to any thing else, than a general weakness, arising from causes that must in a short time, without some remedy or relief, produce their total failure. We did not neglect, however, to patch up and mend them, according to our means, whenever we could discover a broken or weak part. We this day found ourselves surrounded by a shoal of dolphins; some, or one of which, we tried in vain a long time to take. We made a small line from some rigging that was in the boat, fastened on one of the fish-hooks, and tied to it a small piece of white rag; they took not the least notice of it, but continued playing around us, nearly all day, mocking both our miseries and our efforts.

November 30th. This was a remarkably fine day; the weather not exceeded by any that we had experienced since we left the wreck. At one o'clock, I proposed to our boat's crew to kill one of the turtle; two of which we had in our possession. I need not say, that the proposition was hailed with the utmost enthusiasm; hunger had set its ravenous gnawings upon our stomachs, and we waited with impatience to suck the warm flowing blood of the animal. A small fire was kindled in the shell of the turtle, and after dividing the blood (of which there was about a gill), among those of us who felt disposed to drink it, we cooked the remainder, entrails and all, and enjoyed from it an unspeakably fine repast. The stomachs of two or three revolted at the sight of the blood, and refused to partake of it; not even the outrageous thirst that was upon them could induce them to taste it; for myself, I took it like a medicine, to relieve the extreme dryness of my palate, and stopped not to inquire whether it was any thing else than a liquid. After this, I may say exquisite banquet, our bodies were considerably recruited, and I felt my spirits now much higher than they had been at any time before. By observation, this day we found ourselves in latitude 7° 53' S. our distance from the wreck, as nearly as we could calculate, was then about four hundred and eighty miles.

December 1st. From the 1st to the 3rd of December, exclusive, there was nothing transpired of any moment. Our boats as yet kept admirably well together, and the weather was distinguished for its mildness and salubrity. We gathered consolation too from a favourable slant which the wind took to NE, and our situation was not at that moment, we thought, so comfortless as we had been led at first to consider it; but, in our extravagant felicitations upon the blessing of the wind and weather, we forgot our leaks, our weak boats, our own debility, our immense distance from land, the smallness of our stock of provisions; all which, when brought to mind, with the force which they deserved, were too well calculated to dishearten us, and cause us to sigh for the hardships of our lot. Up to the 3rd of December, the raging thirst of our mouths had not been but in a small degree alleviated; had it not been for the pains which that gave us, we should have tasted, during this spell of fine weather, a species of enjoyment, derived from a momentary forgetfulness of our actual situation.

December 3rd. With great joy we hailed the last crumb of our damaged bread, and commenced this day to take our allowance of healthy provisions. The salutary and agreeable effects of this change were felt at first in so slight a degree, as to give us no great cause of comfort or satisfaction; but gradually, as we partook of our small allowance of water, the moisture began to collect in our mouths, and the parching fever of the palate imperceptibly left it. An accident here happened to us which gave us a great momentary spell of uneasiness. The night was dark, and the sky was completely overcast, so that we could scarcely discern each other's boats, when at about ten o'clock, that of the second mate was suddenly missing. I felt for a moment considerable alarm at her unexpected disappearance, but after a little reflection I immediately hove to, struck a light as expeditiously as possible, and hoisted it at the masthead, in a lantern. Our eyes were now directed over every part of the ocean, in search of her, when, to our great joy, we discerned an answering light, about a quarter of a mile to leeward of us; we ran down to it, and it proved to be the lost boat. Strange as the extraordinary interest which we felt in each other's company may appear, and much as our repugnance to separation may seem to imply of weakness, it was the subject of our continual hopes and fears. It is truly remarked, that misfortune more than anything else serves to endear us to our companions. So strongly was this sentiment engrafted upon our feelings, and so closely were the destinies of all of us involuntarily linked together, that, had one of the boats been wrecked, and wholly lost, with all her provisions and water, we should have felt ourselves constrained, by every tie of humanity, to have taken the surviving sufferers into the other boats, and shared our bread and water with them, while a crumb of one or a drop of the other remained. Hard, indeed, would the case have been for all, and much as I have since reflected on the subject, I have not been able to realize, had it so happened, that a sense of our necessities would have allowed us to give so magnanimous and devoted a character to our feelings. I can only speak of the impressions which I recollect I had at the time. Subsequently, however as our situation became more straightened and desperate, our conversation on this subject took a different turn, and it appeared to be a universal sentiment, that such a course of conduct was calculated to weaken the chances of a final deliverance for some, and might be the only means of consigning every soul of us to a horrid death of starvation. There is no question but that an immediate separation, therefore, was the most politic measure that could be adopted, and that every boat should take its own separate chance: while we remained together, should any accident happen, of the nature alluded to, no other course could be adopted, than that of taking the survivors into the other boats, and giving up voluntarily, what we were satisfied could alone prolong our hopes, and multiply the chances of our safety, or unconcernedly witness their struggles in death, perhaps beat them from our boats, with weapons, back into the ocean. The expectation of reaching the land was founded upon a reasonable calculation of the distance, the means, and the subsistence; all which were scanty enough, God knows, and ill adapted to the probable exigencies of the voyage. Any addition to our own demands, in this respect, would not only injure, but actually destroy the whole system which we had laid down, and reduce us to a slight hope, derived either from the speedy death of some of our crew, or the falling in with some vessel. With all this, however, there was a desperate instinct that bound us together; we could not reason on the subject with any degree of satisfaction to our minds, yet we continued to cling to each other with a strong and involuntary impulse. This, indeed, was a matter of no small difficulty, and it constituted, more than any thing else, a source of continual watching and inquietude. We would but turn our eyes away for a few moments, during some dark nights, and presently, one of the boats would be missing. There was no other remedy than to heave to immediately and set a light, by which the missing boat might be directed to us. These proceedings necessarily interfered very much with our speed, and consequently lessened our hopes, but we preferred to submit to it, while the consequences were not so immediately felt, rather than part with the consolation which each other's presence afforded. Nothing of importance took place on the 4th of December; and on the 5th, at night, owing to the extreme darkness, and a strong wind, I again separated from the other boats. Finding they were not to be seen in any direction, I loaded my pistol and fired it twice; soon after the second discharge they made their appearance a short distance to windward, and we joined company, and again kept on our course, in which we continued without any remarkable occurrence, through the 6th and 7th of December. The wind during this period blew very strong, and much more unfavourably. Our boats continued to leak, and to take in a good deal of water over the gunwales.

December 8th. In the afternoon of this day the wind set in E.SE. and began to blow much harder than we had yet experienced it; by twelve o'clock at night it had increased to a perfect gale, with heavy showers of rain, and we now began, from these dreadful indications, to prepare ourselves for destruction. We continued to take in sail by degrees, as the tempest gradually increased, until at last we were obliged to take down our masts. At this juncture we gave up entirely to the mercy of the waves. The sea and rain had wet us to the skin, and we sat down, silently, and with sullen resignation, awaited our fate. We made an effort to catch some fresh water by spreading one of the sails, but after having spent a long time, and obtained but a small quantity in a bucket, it proved to be quite as salt as that from the ocean: this we attributed to its having passed through the sail which had been so often wet by the sea, and upon which, after drying so frequently in the sun, concretions of salt had been formed. It was a dreadful night—cut off from any imaginary relief—nothing remained but to await the approaching issue with firmness and resignation. The appearance of the heavens was dark and dreary, and the blackness that was spread over the face of the waters dismal beyond description. The heavy squalls, that followed each other in quick succession, were preceded by sharp flashes of lightning, that appeared to wrap our little barge in flames. The sea rose to a fearful height, and every wave that came looked as if it must be the last that would be necessary for our destruction.

To an overruling Providence alone must be attributed our salvation from the horrors of that terrible night. It can be accounted for in no other way: that a speck of substance, like that which we were, before the driving terrors of the tempest, could have been conducted safely through it. At twelve o'clock it began to abate a little in intervals of two or three minutes, during which we would venture to raise up our heads and look to windward. Our boat was completely unmanageable; without sails, mast, or rudder, and had been driven, in the course of the afternoon and night, we knew no whither, nor how far. When the gale had in some measure subsided we made efforts to get a little sail upon her, and put her head towards the course we had been steering. My companions had not slept any during the whole night and were dispirited and broken down to such a degree as to appear to want some more powerful stimulus than the fears of death to enable them to do their duty. By great exertions, however, towards morning we again set a double-reefed mainsail and jib upon her, and began to make tolerable progress on the voyage. An unaccountable good fortune had kept the boats together during all the troubles of the night: and the sun rose and showed the disconsolate faces of our companions once more to each other.

December 9th. By twelve o'clock this day we were enabled to set all sail as usual; but there continued to be a very heavy sea running, which opened the seams of the boats, and increased the leaks to an alarming degree. There was, however, no remedy for this but continual bailing, which had now become to be an extremely irksome and laborious task. By observation we found ourselves in latitude 17° 40' S. At eleven o'clock at night, the captain's boat was unexpectedly found to be missing. After the last accident of this kind we had agreed, if the same should again occur, that, in order to save our time, the other boats should not heave to, as usual, but continue on their course until morning, and thereby save the great detention that must arise from such repeated delays. We, however, concluded on this occasion to make a small effort, which, if it did not immediately prove the means of restoring the lost boat, we would discontinue, and again make sail. Accordingly we hove to for an hour, during which time I fired my pistol twice, and obtaining no tidings of the boat, we stood on our course. When daylight appeared she was to leeward of us, about two miles; upon observing her we immediately ran down, and again joined company.

December 10th. I have omitted to notice the gradual advances which hunger and thirst, for the last six days, had made upon us. As the time had lengthened since our departure from the wreck, and the allowance of provision, making the demands of the appetite daily more and more importunate, they had created in us an almost uncontrollable temptation to violate our resolution, and satisfy, for once, the hard yearnings of nature from our stock; but a little reflection served to convince us of the imprudence and unmanliness of the measure, and it was abandoned with a sort of melancholy effort of satisfaction. I had taken into custody, by common consent, all the provisions and water belonging to the boat, and was determined that no encroachments should be made upon it with my consent; nay, I felt myself bound, by every consideration of duty, by every dictate of sense, of prudence, and discretion, without which, in my situation, all other exertions would have been folly itself, to protect them, at the hazard of my life. For this purpose, I locked up in my chest the whole quantity, and never, for a single moment, closed my eyes without placing some part of my person in contact with the chest; and having loaded my pistol, kept it constantly about me. I should not certainly have put any threats in execution as long as the most distant hopes of reconciliation existed; and was determined, in case the least refractory disposition should be manifested (a thing which I contemplated not unlikely to happen, with a set of starving wretches like ourselves), that I would immediately divide our subsistence into equal proportions, and give each man's share into his own keeping. Then, should any attempt be made upon mine, which I intended to mete out to myself, according to exigencies, I was resolved to make the consequences of it fatal. There was, however, the most upright and obedient behavior in this respect manifested by every man in the boat, and I never had the least opportunity of proving what my conduct would have been on such an occasion. While standing on our course this day we came across a small shoal of flying fish: four of which, in their efforts to avoid us, flew against the mainsail, and dropped into the boat; one, having fell near me, I eagerly snatched up and devoured; the other three were immediately taken by the rest, and eaten alive. For the first time I, on this occasion, felt a disposition to laugh, upon witnessing the ludicrous and almost desperate efforts of my five companions, who each sought to get a fish. They were very small of the kind, and constituted but an extremely delicate mouthful, scales, wings, and all for hungry stomachs like ours. From the eleventh to the thirteenth of December inclusive, our progress was very slow, owing to light winds and calms; and nothing transpired of any moment, except that on the eleventh we killed the only remaining turtle, and enjoyed another luxuriant repast, that invigorated our bodies, and gave a fresh flow to our spirits. The weather was extremely hot, and we were exposed to the full force of a meridian sun, without any covering to shield us from its burning influence, or the least breath of air to cool its parched rays. On the thirteenth day of December we were blessed with a change of wind to the northward, that brought us a most welcome and unlooked for relief. We now, for the first time, actually felt what might be deemed a reasonable hope of our deliverance; and with hearts bounding with satisfaction, and bosoms swelling with joy, we made all sail to the eastward. We imagined we had run out of the trade-winds, and had got into the variables, and should, in all probability, reach the land many days sooner than we expected. But, alas! Our anticipations were but a dream, from which we shortly experienced a cruel awakening. The wind gradually died away, and at night was succeeded by a perfect calm, more oppressive and disheartening to us, from the bright prospects which had attended us during the day. The gloomy reflections that this hard fortune had given birth to, were succeeded by others, of a no less cruel and discouraging nature, when we found the calm continue during the fourteenth, fifteenth, and sixteenth of December inclusive. The extreme oppression of the weather, the sudden and unexpected prostration of our hopes, and the consequent dejection of our spirits, set us again to thinking, and filled our souls with fearful and melancholy forebodings. In this state of affairs, seeing no alternative left us but to employ to the best advantage all human expedients in our power, I proposed, on the fourteenth, to reduce our allowance of provisions one half. No objections were made to this arrangement: all submitted, or seemed to do so, with an admirable fortitude and forbearance. The proportion which our stock of water bore to our bread was not large; and while the weather continued so oppressive, we did not think it advisable to diminish our scanty pittance; indeed, it would have been scarcely possible to have done so, with any regard to our necessities, as our thirst had become now incessantly more intolerable than hunger, and the quantity then allowed was barely sufficient to keep the mouth in a state of moisture, for about one-third of the time. "Patience and long-suffering" was the constant language of our lips: and a determination, strong as the resolves of the soul could make it, to cling to existence as long as hope and breath remained in us. In vain was every expedient tried to relieve the raging fever of the throat by drinking salt water, and holding small quantities of it in the mouth, until, by that means, the thirst was increased to such a degree, as even to drive us to despairing, and vain relief from our own urine. Our sufferings during these calm days almost succeeded human belief. The hot rays of the sun beat down upon us to such a degree, as to oblige us to hang over the gunwale of the boat, into the sea, to cool our weak and fainting bodies. This expedient afforded us, however, a grateful relief, and was productive of a discovery of infinite importance to us. No sooner had one of us got on the outside of the gunwale than he immediately observed the bottom of the boat to be covered with a species of small clam, which, upon being tasted, proved a most delicious and agreeable food. This was no sooner announced to us, than we commenced to tear them off and eat them, for a few minutes, like a set of gluttons; and, after having satisfied the immediate craving of the stomach, we gathered large quantities and laid them up in the boat; but hunger came upon us again in less than half an hour afterwards within which time they had all disappeared. Upon attempting to get in again, we found ourselves so weak as to require each other's assistance; indeed, had it not been for three of our crew, who could not swim, and who did not, therefore, get overboard, I know not by what means we should have been able to have resumed our situations in the boat.

On the fifteenth our boat continued to take in water so fast from her leaks, and the weather proving so moderate, we concluded to search out the bad places, and endeavour to mend them as well as we should be able. After a considerable search, and, removing the ceiling near the bows, we found the principal opening was occasioned by the starting of a plank or strake in the bottom of the boat, next to the keel. To remedy this, it was now absolutely necessary to have access to the bottom. The means of doing which did not immediately occur in our minds. After a moment's reflection, however, one of the crew, Benjamin Lawrence, offered to tie a rope around his body, take a boat's hatchet in his hand, and thus go under the water, and hold the hatchet against a nail, to be driven through from the inside, for the purpose of clenching it. This was, accordingly, all effected, with some little trouble, and answered the purpose much beyond our expectations. Our latitude was this day 21° 42' South. The oppression of our weather still continuing through the sixteenth, bore upon our health and spirits with an amazing force and severity. The most disagreeable excitements were produced by it, which, added to the disconsolate endurance of the calm, called loudly for some mitigating expedient—some sort of relief to our prolonged sufferings. By our observations today we found, in addition to our other calamities, that we had been urged back from our progress, by the heave of the sea, a distance of ten miles; and were still without any prospect of wind. In this distressing posture of our affairs, the captain proposed that we should commence rowing, which, being seconded by all, we immediately concluded to take a double allowance of provision and water for a day, and row, during the cool of the nights, until we should get a breeze from some quarter or other. Accordingly, when night came, we commenced our laborious operations: we made but a very sorry progress. Hunger and thirst, and long inactivity, had so weakened us, that in three hours every man gave out, and we abandoned the further prosecution of the plan. With the sunrise the next morning, on the seventeenth, a light breeze sprung up from the SE. and, although directly ahead, it was welcomed with almost frenzied feelings of gratitude and joy.

December 18th. The wind had increased this day considerably, and by twelve o'clock blew a gale; veering from SE. to E.SE. Again we were compelled to take in all sail, and lie to for the principal part of the day. At night, however, it died away, and the next day, the nineteenth, proved very moderate and pleasant weather, and we again commenced to make a little progress.

December 20th. This was a day of great happiness and joy. After having experienced one of the most distressing nights in the whole catalogue of our sufferings, we awoke to a morning of comparative luxury and pleasure. About 7 o'clock, while we were sitting dispirited, silent, and dejected, in our boats, one of our companions suddenly and loudly called out, "there is land!" We were all aroused in an instant, as if electrified, and casting our eyes to leeward, there indeed, was the blessed vision before us, "as plain and palpable" as could be wished for. A new and extraordinary impulse now took possession of us. We shook off the lethargy of our senses, and seemed to take another, and a fresh existence. One or two of my companions, whose lagging spirits, and worn out frames, had begun to inspire them with an utter indifference to their fate, now immediately brightened up, and manifested a surprising alacrity and earnestness to gain, without delay, the much wished for shore. It appeared at first a low, white beach, and lay like a basking paradise before our longing eyes. It was discovered nearly at the same time by the other boats, and a general burst of joy and congratulation now passed between us. It is not within the scope of human calculation, by a mere listener to the story, to divine what the feelings of our hearts were on this occasion. Alternate expectation, fear, gratitude, surprise, and exultation, each swayed our minds, and quickened our exertions. We ran down for it, and at 11 o'clock a.m. we were within a quarter of a mile of the shore. It was an island, to all appearance, as nearly as we could determine it, about six miles long, and three broad; with a very high, rugged shore, and surrounded by rocks; the sides of the mountains were bare, but on the tops it looked fresh and green with vegetation. Upon examining our navigators, we found it was Ducie's Island, lying in latitude 24° 40' S. longitude 124° 40' W. 14. A short moment sufficed for reflection, and we made immediate arrangements to land. None of us knew whether the island was inhabited or not, nor what it afforded, if any thing; if inhabited, it was uncertain whether by beasts or savages; and a momentary suspense was created, by the dangers which might possibly arise by proceeding without due preparation and care. Hunger and thirst, however, soon determined us, and having taken the musket and pistols, I, with three others, effected a landing upon some sunken rocks, and waded thence to the shore. Upon arriving at the beach, it was necessary to take a little breath, and we laid down for a few minutes to rest our weak bodies, before we could proceed. Let the reader judge, if he can, what must have been our feelings now! Bereft of all comfortable hopes of life, for the space of thirty days of terrible suffering; our bodies wasted to mere skeletons, by hunger and thirst, and death itself staring us in the face; to be suddenly and unexpectedly conducted to a rich banquet of food and drink, which subsequently we enjoyed for a few days, to our full satisfaction; and he will have but a faint idea of the happiness that here fell to our lot. We now, after a few minutes, separated, and went different directions in search of water; the want of which had been our principal privation, and called for immediate relief. I had not proceeded far in my excursion, before I discovered a fish, about a foot and a half in length, swimming along in the water close to the shore. I commenced an attack upon him with the breach of my gun, and struck him, I believe, once, and he ran under a small rock, that lay near the shore, from whence I took him with the aid of my ramrod, and brought him up on the beach, and immediately fell to eating. My companions soon joined in the repast; and in less than ten minutes the whole was consumed, bones, and skin, and scales, and all. With full stomachs, we imagined we could now attempt the mountains, where, if in any part of the island, we considered water would be most probably obtained.

I accordingly clambered, with excessive labour, suffering, and pain, up amongst the bushes, roots, and underwood, of one of the crags, looking in all directions in vain, for every appearance of water that might present itself. There was no indication of the least moisture to be found, within the distance to which I had ascended, although my strength did not enable me to get higher than about 20 feet. I was sitting down at the height that I had attained, to gather a little breath, and ruminating upon the fruitlessness of my search, and the consequent evils and continuation of suffering that is necessarily implied, when I perceived that the tide had risen considerably since our landing, and threatened to cut off our retreat to the rocks, by which alone we should be able to regain our boats. I therefore determined to proceed again to the shore, and inform the captain and the rest of our want of success in procuring water, and consult upon the propriety of remaining at the island any longer. I never for one moment lost sight of the main chance, which I conceived we still had, of either getting to the coast, or of meeting with some vessel at sea; and felt that every minute's detention, without some equivalent object, was lessening those chances, by a consumption of the means of our support. When I had got down, one of my companions informed me, that he had found a place in a rock some distance off, from which the water exuded in small drops, at intervals of about five minutes, that he had, by applying his lips to the rock, obtained a few of them, which only served to whet his appetite, and from which nothing like the least satisfaction had proceeded.

I immediately resolved in my own mind, upon this information, to advise remaining until morning, to endeavour to make a more thorough search the next day, and with our hatchets to pick away the rock which had been discovered, with the view of increasing, if possible, the run of the water. We all repaired again to our boats, and there found that the captain had the same impressions as to the propriety of our delay until morning. We therefore landed; and having hauled our boats up on the beach, laid down in them that night, free from all the anxieties of watching and labour, and amid all our sufferings, gave ourselves up to an unreserved forgetfulness and peace of mind, that seemed so well to accord with the pleasing anticipations that this day had brought forth. It was but a short space, however, until the morning broke upon us; and sense, and feeling, and gnawing hunger, and the raging fever of thirst then redoubled my wishes and efforts to explore the island again. We had obtained, that night, a few crabs, by traversing the shore a considerable distance, and a few very small fish; but waited until the next day, for the labours of which, we considered a night of refreshing and undisturbed repose would better qualify us.

December 21st. We had still reserved our common allowance, but it was entirely inadequate for the purpose of supplying the raging demand of the palate, and such an excessive and cruel thirst was created, as almost to deprive us of the power of speech. The lips became cracked and swollen, and a sort of glutinous saliva collected in the mouth, disagreeable to the taste, and intolerable beyond expression. Our bodies had wasted away to almost skin and bone, and possessed so little strength, as often to require each other's assistance in performing some of its weakest functions. Relief, we now felt, must come soon, or nature would sink. The most perfect discipline was still maintained, in respect to our provisions; and it now became our whole object, if we should not be able to replenish our subsistence from the island, to obtain, by some means or other, a sufficient refreshment to enable us to prosecute our voyage.

Our search for water accordingly again commenced with the morning; each of us took a different direction, and prosecuted the examination of every place where there was the least indication of it; the small leaves of the shrubbery, affording a temporary alleviation, by being chewed in the mouth, and but for the peculiarly bitter taste which those of the island possessed, would have been an extremely grateful substitute. In the course of our rambles too, along the sides of the mountain, we would now and then meet with tropic birds, of a beautiful figure and plumage occupying small holes in the sides of it, from which we plucked them without the least difficulty. Upon our approaching them they made no attempts to fly, nor did they appear to notice us at all. These birds served us for a fine repast; numbers of which were caught in the course of the day, cooked by fires which we made on the shore, and eaten with the utmost avidity. We found also a plant, in taste not unlike the peppergrass, growing in considerable abundance in the crevices of the rocks, and which proved to us a very agreeable food, by being chewed with the meat of the birds. These, with birds' nests, some of them full of young, and others of eggs, a few of which we found in the course of the day, served us for food, and supplied the place of our bread; from the use of which, during our stay here, we had restricted ourselves. But water, the great object of all our anxieties and exertions, was nowhere to be found, and we began to despair of meeting with it on the island. Our state of extreme weakness, and many of us without shoes or any covering for the feet, prevented us from exploring any great distance, lest by some sudden faintness, or over exertion, we should not be able to return, and at night be exposed to attacks of wild beasts, which might inhabit the island, and be alike incapable of resistance, as beyond the reach of the feeble assistance that otherwise could be afforded to each. The whole day was thus consumed in picking up whatever had the least shape or quality of sustenance, and another night of misery was before us, to be passed without a drop of water to cool our parching tongues. In this state of affairs, we could not reconcile it to ourselves to remain longer at this place; a day, and hour, lost to us unnecessarily here, might cost us our preservation. A drop of the water that we then had in our possession might prove, in the last stages of our debility, the very cordial of life. I addressed the substance of these few reflections to the captain, who agreed with me in opinion, upon the necessity of taking some decisive steps in our present dilemma. After some considerable conversation on this subject, it was finally concluded, to spend the succeeding day in the further search for water, and if none should be found, to quit the island the morning after.

December 23rd. We had been employed during the last night to various occupations, according to the feelings or the wants of the men; some continued to wander about the shore, and to short distances in the mountains, still seeking for food and water; others hung about the beach, near the edge of the sea, endeavouring to take the little fish that came about them. Some slept, insensible to every feeling but rest; while others spent the night in talking of their situation, and reasoning upon the probabilities of their deliverance. The dawn of day aroused us again to labour, and each of us pursued his own inclination, as the course taken over the island after water. My principal hope was founded upon my success in picking the rocks where the moisture had been discovered the day before, the thither I hastened as soon as my strength would enable me to get there. It was about a quarter of a mile from what

I may call our encampment; and with two men, who had accompanied me. I commenced my labours with a hatchet and an old chisel. The rock proved to be very soft, and in a very short time I had obtained a considerable hole, but, alas! Without the least wished-for effect. I watched it for some little time with great anxiety, hoping that, as I increased the depth of the hole, the water would presently flow; but all my hopes and efforts were unavailing, and at last I desisted from further labour, and sat down near it in utter despair. As I turned my eyes towards the beach I saw some of the men in the act of carrying a keg along from the boats, with, I thought, an extraordinary spirit and activity; and the idea suddenly darted across my mind that they had found water, and were taking a keg to fill it. I quitted my seat in a moment, made the best of my way towards them, with a palpitating heart, and before I came up with them, they gave me the cheering news that they had found a spring of water. I felt, at that moment, as if I could have fallen down and thanked God for this signal act of his mercy. The sensation that I experienced was indeed strange, and such as I shall never forget. At one instant I felt an almost choking excess of joy, and at the next I wanted the relief of a flood of tears. When I arrived at the spot, whither I had hastened as fast as my weak legs would carry me, I found my companions had all taken their fill, and with an extreme degree of forbearance I then satisfied myself, by drinking in small quantities, and at intervals of two or three minutes apart. Many had, notwithstanding the remonstrances of prudence, and, in some cases, force, laid down and thoughtlessly swallowed large quantities of it, until they could drink no more. The effect of this was, however, neither so sudden nor bad as we had imagined; it only served to make them a little stupid and indolent for the remainder of the day.

Upon examining the place from whence we had obtained this miraculous and unexpected succor, we were equally astonished and delighted with the discovery. It was on the shore, above which the sea flowed to the depth of near six feet; and we could procure the water, therefore, from it only when the tide was down. The crevice from which it rose was in a flat rock, large surfaces of which were spread around, and composed the face of the beach. We filled our two kegs before the tide rose, and went back again to our boats. The remainder of this day was spent in seeking for fish, crabs, birds, and any thing else that fell in our way, that could contribute to satisfy our appetites; and we enjoyed, during that night, a most comfortable and delicious sleep, unattended with those violent cravings of hunger and thirst, that had poisoned our slumbers for so many previous ones. Since the discovery of the water, too, we began to entertain different notions altogether of our situation. There was no doubt we might here depend upon a constant and ample supply of it as long as we chose to remain, and, in all probability, we could manage to obtain food, until the island should be visited by some vessel, or time allowed to devise other means of leaving it. Our boats would still remain to us: a stay here might enable us to mend, straighten, and put them in more perfect order for the sea, and get ourselves so far recruited as to be able to endure, if necessary, a more protracted voyage to the main land. I made a silent determination in my own mind that I would myself pursue something like this plan, whatever might be the opinion of the rest; but I found no difference in the views of any of us as to this matter. We, therefore, concluded to remain at least four or five days, within which time it could be sufficiently known whether it would be adviseable [*sic*] to make any arrangement for a more permanent abode.

December 23rd. At 11 o'clock a.m. we again visited our spring: the tide had fallen to about a foot below it, and we were able to procure, before it rose again, about twenty gallons of water. It was at first a little brackish, but soon became fresh, from the constant supply from the rock, and the departure of the sea. Our observations this morning tended to give us every confidence in its quantity and quality, and we, therefore, rested perfectly easy in our minds on the subject, and commenced to make further discoveries about the island. Each man sought for his own daily living, on whatsoever the mountains, the shore, or the sea, could furnish him with; and every day, during our stay there, the whole time was employed in roving about for food. We found, however, on the twenty-fourth, that we had picked up, on the island, every thing that could be got at, in the way of sustenance; and, much to our surprise, some of the men came in at night and complained of not having gotten sufficient during the day to satisfy the cravings of their stomachs. Every accessible part of the mountain, contiguous to us, or within the reach of our weak enterprise, was already ransacked, for birds' eggs and grass, and was rifled of all that they contained: so that we began to entertain serious apprehensions that we should not be able to live long here; at any rate, with the view of being prepared, as well as possible, should necessity at any time oblige us to quit it, we commenced, on the twenty-fourth, to repair our boats, and continued to work upon them all that and the succeeding day. We were enabled to do this, with much facility, by drawing them up and turning them over on the beach, working by spells of two or three hours at a time, and then leaving off to seek for food. We procured our water daily, when the tide would leave the shore: but on the evening of the twenty-fifth, found that a fruitless search for nourishment had not repaid us for the labours of a whole day. There was no one thing on the island upon which we could in the least degree rely, except the peppergrass, and of that the supply was precarious, and not much relished without some other food. Our situation here, therefore, now became worse than it would have been in our boats on the ocean; because, in the latter case, we should be still making some progress towards the land, while our provisions lasted, and the chance of falling in with some vessel be considerably increased. It was certain that we ought not to remain here unless upon the strongest assurances in our own minds, of sufficient sustenance, and that, too, in regular supplies, that might be depended upon. After much conversation amongst us on this subject, and again examining our navigators, it was finally concluded to set sail for Easter Island, which we found to be E.SE. from us in latitude 27° 9' S. longitude 109° 35' W. All we knew of this island was, that it existed as laid down in the books; but of its extent, productions, or inhabitants, if any, we were entirely ignorant; at any rate, it was nearer by eight hundred and fifty miles to the coast, and could not be worse in its productions than the one we were about leaving.

The twenty-sixth of December was wholly employed in preparations for our departure; our boats were hauled down to the vicinity of the spring and our casks, and every thing else that would contain it, filled with water.

There had been considerable talk between three of our companions, about their remaining on this island, and taking their chance both for a living, and an escape from it; and as the time drew near at which we were to leave, they made up their minds to stay behind. The rest of us could make no objection to their plan, as it lessened the load of our boats, allowed us their share of the provisions, and the probability of their being able to sustain themselves on the island was much stronger than that of our reaching the main land. Should we, however, ever arrive safely, it would become our duty, and we so assured them, to give information of their situation, and make every effort to procure their removal from thence; which we accordingly afterwards did.

Their names were William Wright of Barnstable, Massachusetts, Thomas Chapple of Plymouth, England, and Seth Weeks of the former place. They had begun, before we came away, to construct a sort of habitation, composed of the branches of trees, and we left with them every article that could be spared from the boats. It was their intention to build a considerable dwelling, that could protect them from the rains, as soon as time and materials could be provided. The captain wrote letters, to be left on the island, giving information of the fate of the ship, and that of our own; and stating that we had set out to reach Easter Island, with further particulars, intended to give notice (should our fellow-sufferers die there, and the place be ever visited by any vessel) of our misfortunes. These letters were put in a tin case, enclosed in a small wooden box, and nailed to a tree, on the west side of the island, near our landing place. We had observed, some days previously, the name of a ship, "The Elizabeth", cut out in the bark of this tree, which rendered it indubitable that one of that name had once touched here. There was, however, no date to it, or any thing else, by which any further particulars could be made out.

December 27th. I went, before we set sail this morning, and procured for each boat a flat stone, and two arms-full of wood, with which to make a fire in our boats, should it become afterwards necessary in the further prosecution of our voyage; as we calculated we might catch a fish, or a bird, and in that case be provided with the means of cooking it; otherwise, from the intense heat of weather, we knew they could not be preserved from spoiling. At ten o'clock a.m. the tide having risen far enough to allow our boats to float over the rocks, we made all sail, and steered around the island, for the purpose of making a little further observation, which would not detail us any time, and might be productive of some unexpected good fortune. Before we started we missed our three companions, and found they had not come down, either to assist us to get off, nor to take any kind of leave of us.

I walked up the beach towards their rude dwelling, and informed them that we were then about to set sail, and should probably never see them more. They seemed to be very much affected, and one of them shed tears. They wished us to write to their relations, should Providence safely direct us again to our homes, and said but little else. They had every confidence in being able to procure a subsistence there as long as they remained: and, finding them ill at heart about taking any leave of us, I hastily bid them "good-bye," hoped they would do well, and came away. They followed me with their eyes until I was out of sight, and I never saw more of them.

On the NW side of the island we perceived a fine white beach, on which we imagined we might land, and in a short time ascertain if any further useful discoveries could be effected, or any addition made to our stock of provisions; and having set ashore five or six of the men for this purpose, the rest of us shoved off the boats and commenced fishing. We saw a number of sharks, but all efforts to take them proved ineffectual; and we got but a few small fish, about the size of a mackerel, which we divided amongst us. In this business we were occupied for the remainder of the day, until six o'clock in the afternoon, when the men, having returned to the shore from their search in the mountains, brought a few birds, and we again set sail and steered directly for Easter Island. During that night, after we had got quite clear of the land, we had a fine strong breeze from the NW.; we kept our fires going, and cooked our fish and birds, and felt our situation as comfortable as could be expected. We continued on our course, consuming our provisions and water as sparingly as possible, without any material incident, until the thirtieth, when the wind hauled out E.SE. directly ahead, and so continued until the thirty-first, when it again came to the northward, and we resumed our course.

On the third of January we experienced heavy squalls from the W.SW. accompanied with dreadful thunder and lightning, that threw a gloomy and cheerless aspect over the ocean, and incited a recurrence of some of those heavy and desponding moments that we had before experienced. We commenced from Ducies Island [*sic*] to keep a regular reckoning, by which, on the fourth of January, we found we had got to the southward of Easter Island, and the wind prevailing E.NE. we should not be able to get on the eastward, so as to reach it. Our birds and fish were all now consumed, and we had begun again upon our short allowance of bread. It was necessary, in this state of things, to change our determination of going to Easter Island, and shape our course in some other direction, where the wind would allow of our going. We had but little hesitation in concluding, therefore, to steer for the island of Juan Fernandez, which lay about E.SE. from us, distant two thousand five hundred miles. We bent our course accordingly towards it, having for the two succeeding days very light winds, and suffering excessively from the intense heat of the sun. The seventh brought us a change of wind in the northward, and at twelve o'clock we found ourselves in latitude 30° 18' S. longitude 117° 29' W. We continued to make what progress we could to the eastward.

January 10th. Matthew P. Joy, the second mate, had suffered from debility, and the privations we had experienced, much beyond any of the rest of us, and was on the eighth removed to the captain's boat, under the impression that he would be more comfortable there, and more attention and pains be bestowed in nursing and endeavouring to comfort him. This day being calm, he manifested a desire to be taken back again; but at 4 o'clock in the afternoon, after having been, according to his wishes, placed in his own boat, he died very suddenly after his removal. On the eleventh, at six o'clock in the morning, we sewed him up in his clothes, tied a large stone to his feet, and having brought all the boats to, consigned him in a solemn manner to the ocean. This man did not die of absolute starvation, although his end was no doubt very much hastened by his sufferings. He had a weak and sickly constitution, and complained of being unwell the whole voyage. It was an incident, however, which threw a gloom over our feelings for many days. In consequence of his death, one man from the captain's boat was placed in that from which he died to supply his place, and we stood away again from our course.

On the 12th of January we had the wind from the NW., which commenced in the morning, and came on to blow before night a perfect gale. We were obliged to take in all sail and run before the wind. Flashes of lightning were quick and vivid, and the rain came down in cataracts. As however that gale blew us fairly on our course, and our speed being great during the day, we derived, I may say, even pleasure from the uncomfortableness and fury of the storm. We were apprehensive that in the darkness of this night we should be separated, and made arrangements, each boat to keep an E.SE course all night. About eleven o'clock my boat being ahead a short distance of the others, I turned my head back, as I was in the habit of doing every minute, and neither of the others were to be seen. It was blowing and raining at this time as if the heavens were separating, and I knew not hardly at the moment what to do. I hove my boat to the wind, and lay drifting about an hour, expecting every moment that they would come up with me, but not seeing anything of them, I put away again, and stood on the course agreed upon, with strong hopes that daylight would enable me to discover them again. When the morning dawned, in vain did we look over every part of the ocean for our companions; they were gone! And we saw no more of them afterwards. It was folly to repine at the circumstances; it could neither be remedied, nor could sorrow secure their return; but it was impossible to prevent ourselves feeling all poignancy and bitterness that characterizes the separation of men who have long suffered in each other's company, and whose interests and feelings fate had so closely linked together. By our observation, we separated in latitude 32° 16' S. longitude 112° 20' W. For many days after this accident, our progress was attended with dull and melancholy reflections. We had lost the cheering of each other's faces, that, which strange as it is, we so much required in both our mental and bodily distresses. The 14th of January proved another very squally and rainy day. We had now been nineteen days from the island, and had made a distance of about 900 miles: necessity began to whisper us, that a still further reduction of our allowance must take place, or we must abandon altogether the hopes of reaching the land, and rely wholly on the chance of being taken up by a vessel. But how to reduce the daily quantity of food, with any regard to life itself, was a question of the utmost consequence. Upon our first leaving the wreck, the demands of the stomach had been circumscribed to the smallest possible compass; and subsequently before reaching the island, a diminution had taken place of nearly one-half; and it was now, from a reasonable calculation, become necessary even to curtail that at least one-half; which must, in a short time, reduce us to mere skeletons again. We had a full allowance of water, but it only served to contribute to our debility; our bodies deriving but the scanty support which an ounce and a half of bread for each man afforded. It required a great effort to bring matters to this dreadful alternative, either to feed our bodies and our hopes a little longer, or in the agonies of hunger to seize upon and devour our provisions, and coolly await the approach of death.

We were as yet, just able to move about in our boats, and slowly perform the necessary labours appertaining to her; but we were fast wasting away with the relaxing effects of the water, and we daily almost perished under the torrid rays of a meridian sun; to escape which, we would lie down in the bottom of the boat, cover ourselves over with the sails, and abandon her to the mercy of the waves. Upon attempting to rise again, the blood would rush into the head, and an intoxicating blindness come over us, almost to occasion our suddenly falling down again. A slight interest was still kept up in our minds by the distant hopes of yet meeting with the other boats, but it was never realized. An accident occurred at night, which gave me a great cause of uneasiness, and led me to an unpleasant rumination upon the probable consequences of a repetition of it. I had laid down in the boat without taking the usual precaution of securing the lid of the provision-chest, as I was accustomed to do, when one of the white men awoke me, and informed me that one of the blacks had taken some bread from it. I felt at the moment the highest indignation and resentment at such conduct in any of our crew, and immediately took my pistol in my hand, and charged him if he had taken any, to give it up without the least hesitation, or I should instantly shoot him!—He became at once very much alarmed, and, trembling, confessed the fact, pleading the hard necessity that urged him to it: he appeared to be very penitent for his crime, and earnestly swore that he would never be guilty of it again. I could not find it in my soul to extend towards him the least severity on this account, however much, according to the strict imposition which we felt upon ourselves it might demand it. This was the first infraction; and the security of our lives, our hopes of redemption from our sufferings, loudly called for a prompt and signal punishment; but every humane feeling of nature plead in his behalf, and he was permitted to escape, with the solemn injunction, that a repetition of the same offence would cost him his life.

I had almost determined upon this occurrence to divide our provisions, and give to each man his share of the whole stock; and should have done so in the height of my resentment, had it not been for the reflection that some might, by imprudence, be tempted to go beyond the daily allowance, or consume it all at once, and bring on a premature weakness or starvation: this would of course disable them for the duties of the boat, and reduce our chances of safety and deliverance.

On the 15th of January, at night, a very large shark was observed swimming about us in a most ravenous manner, making attempts every now and then upon different parts of the boat, as if he would devour the very wood with hunger; he came several times and snapped at the steering oar, and even the stern-post. We tried in vain to stab him with a lance, but were so weak as not to be able to make any impression upon his hard skin; he was so much larger than an ordinary one, and manifested such a fearless malignity, as to make us afraid of him; and our utmost efforts, which were at first directed to kill him for prey, became in the end self-defense. Baffled however in all his hungry attempts upon us, he shortly made off.

On the 16th of January, we were surrounded with porpoises in great numbers, that followed us nearly an hour, and which also defied all manoeuvres to catch them. The 17th and 18th proved to be calm; and the distresses of a cheerless prospect and a burning hot sun, were again visited upon our devoted heads.

We began to think that Divine Providence had abandoned us at last; and it was but an unavailing effort to endeavour to prolong a now tedious existence. Horrible were the feelings that took possession of us!—The contemplation of a death of agony and torment, refined by the most dreadful and distressing reflections, absolutely prostrated both body and soul. There was not a hope now remaining to us but that which was derived from a sense of the mercies of our creator. The night of the 18th was a despairing era in our sufferings; our minds were wrought up to the highest pitch of dread and apprehension for our fate, and all in them was dark, gloomy, and confused. About 8 o'clock, the terrible noise of whale-spouts near us sounded in our ears: we could distinctly hear the furious thrashing of their tails in the water, and our weak minds pictured out their appalling and hideous aspects. One of my companions, the black man, took an immediate fright, and solicited me to take out the oars, and endeavour to get away from them. I consented to his using any means for that purpose; but alas! It was wholly out of our power to raise a single arm in our own defense. Two or three of the whales came down near us, and went swiftly off across our stern, blowing and spouting at a terrible rate; they, however, after an hour or two disappeared, and we saw no more of them. The next day, the 19th of January, we had extremely boisterous weather, with rain, heavy thunder and lightning, which reduced us again to the necessity of taking in all sail and lying to. The wind blew from every point of the compass within the twenty-four hours, and at last towards the next morning settled at E.NE. a strong breeze.

January 20. The black man, Richard Peterson, manifested today symptoms of a speedy dissolution; he had been lying between the seats in the boat, utterly dispirited and broken down, without being able to do the least duty, or hardly to place his hand to his head for the last three days, and had this morning made up his mind to die rather than endure further misery: he refused his allowance; said he was sensible of his approaching end, and was perfectly ready to die: in a few minutes he became speechless, the breath appeared to be leaving his body without producing the least pain, and at four o'clock he was gone. I had two days previously, conversations with him on the subject of religion, on which he reasoned very sensibly, and with much composure; and begged me to let his wife know his fate, if ever I reached home in safety. The next morning we committed him to the sea, in latitude 35° 07' S. longitude 105° 46' W. The wind prevailed to the eastward until the 24th of January, when it again fell calm. We were now in a most wretched and sinking state of debility, hardly able to crawl around the boat, and possessing but strength enough to convey our scanty morsel to our mouths. When I perceived this morning that it was calm, my fortitude almost forsook me. I thought to suffer another scorching day, like the last we had experienced, would close before night the scene of our miseries; and I felt many a despairing moment that day, that had well nigh proved fatal. It required an effort to look calmly forward, and contemplate what was yet in store for us, beyond what I felt I was capable of making; and what it was that buoyed me above all the terrors which surrounded us, God alone knows. Our ounce and a half of bread, which was to serve us all day, was in some cases greedily devoured, as if life was to continue but another moment; and at other times, it was hoarded up and eaten crumb by crumb, at regular intervals during the day, as if it was to last us for ever. To add to our calamities, biles began to break out upon us, and our imaginations shortly became as diseased as our bodies. I laid down at night to catch a few moments of oblivious sleep, and immediately my starving fancy was at work. I dreamt of being placed near a splendid and rich repast, where there was every thing that the most dainty appetite could desire; and of contemplating the moment in which we were to commence to eat with enraptured feelings of delight; and just as I was about to partake of it, I suddenly awoke to the cold realities of my miserable situation. Nothing could have oppressed me so much. It set such a longing frenzy for victuals in my mind, that I felt as if I could have wished the dream to continue for ever, that I never might have awoke from it. I cast a sort of vacant stare about the boat, until my eyes rested upon a bit of tough cow-hide, which was fastened to one of the oars; I eagerly seized and commenced to chew it, but there was no substance in it, and it only served to fatigue my weak jaws, and add to my bodily pains. My fellow sufferers murmured very much the whole time and continued to press me continually with questions upon the probability of our reaching land again. I kept constantly rallying my spirits to enable me to afford them comfort. I encouraged them to bear up against all evils, and if we must perish, to die in our own cause, and not weakly distrust the providence of the Almighty, by giving ourselves up to despair. I reasoned with them, and told them that we would not die sooner by keeping up our hopes; that the dreadful sacrifices and privations we endured were to preserve us from death, and were not to be put in competition with the price which we set upon our lives, and their value to our families; it was, besides, unmanly, to repine at what neither admitted of alleviation nor cure; and withal, that it was our solemn duty to recognize in our calamities an overruling divinity, by whose mercy we might be suddenly snatched from peril, and to rely upon him alone, "Who tempers the wind to the shorn lamb."

The three following days, the 25th, 26th, and 27th, were not distinguished by any particular circumstances. The wind still prevailed to the eastward, and by its obduracy, almost tore the very hopes of our hearts away: it was impossible to silence the rebellious repinings of our nature, at witnessing such a succession of hard fortune against us. It was our cruel lot not to have had one bright anticipation realized—not one wish of our thirsting souls gratified. We had, at the end of these three days, been urged to the southward as far as latitude 36°; into a chilly region, where rains and squalls prevailed; and we now calculated to tack and stand back to the northward: after much labour, we got our boat about; and so great was the fatigue attending this small exertion of our bodies, that we all gave up for a moment and abandoned her to her own course.—Not one of us had now strength sufficient to steer, or indeed to make one single effort towards getting the sails properly trimmed, to enable us to make any headway. After an hour or two or relaxation, during which the horrors of our situation came upon us with a despairing force and effect, we made a sudden effort and got our sails into such a disposition, as that the boat would steer herself; and we then threw ourselves down awaiting the issue of time to bring us relief, or to take us from the scene of our troubles. We could now do nothing more; strength and spirits were totally gone; and what indeed could have been the narrow hopes, that in our situation, then bound us to life?

January 28. Our spirits this morning were hardly sufficient to allow of our enjoying a change of the wind, which took place to the westward.—It had nearly become indifferent to us from what quarter it blew: nothing but the slight chance of meeting with a vessel remained to us now: it was this narrow comfort alone, that prevented me from lying down at once to die. But fourteen days' stinted allowance of provisions remained, and it was absolutely necessary to increase the quantity to enable us to live five days longer: we therefore partook of it, as pinching necessity demanded, and gave ourselves wholly up to the guidance and disposal of our Creator.

The 29th and 30th of January, the wind continued west, and we made considerable progress until the 31st, when it again came ahead, and prostrated all our hopes. On the 1st of February, it changed again to the westward, and on the 2nd and 3rd blew to the eastward; and we had it light and variable until the 8th of February. Our sufferings were now drawing to a close; a terrible death appeared shortly to await us; hunger became violent and outrageous, and we prepared for a speedy release from our troubles; our speech and reason were both considerably impaired, and we were reduced to be at this time, certainly the most helpless and wretched of the whole human race. Isaac Cole, one of our crew, had the day before this, in a fit of despair, thrown himself down in the boat, and was determined there calmly to wait for death. It was obvious that he had no chance; all was dark he said in his mind, not a single ray of hope was left for him to dwell upon; and it was folly and madness to be struggling against what appeared so palpably to be our fixed and settled destiny. I remonstrated with him as effectually as the weakness both of my body and understanding would allow of; and what I said appeared for a moment to have a considerable effect: he made a powerful and sudden effort, half rose up, crawled forward and hoisted the jib, and firmly and loudly cried that he would not give up; that he would live as long as the rest of us—but alas! this effort was but the hectic fever of the moment, and he shortly again relapsed into a state of melancholy and despair. This day his reason was attacked, and he became about 9 o'clock in the morning a most miserable spectacle of madness: he spoke coherently about every thing, calling loudly for a napkin and water, and then lying stupidly and senselessly down in the boat again, would close his hollow eyes, as if in death. About 10 o'clock, we suddenly perceived that he became speechless; we got him as well as we were able upon a board, placed on one of the seats of the boat, and covering him up with some old clothes, left him to his fate. He lay in the greatest pain and apparent misery, groaning piteously until four o'clock, when he died, in the most horrid and frightful convulsions I ever witnessed. We kept his corpse all night, and in the morning my two companions began as of course to make preparations to dispose of it in the sea; when after reflecting on the subject all night, I addressed them on the painful subject of keeping the body for food!! Our provisions could not possibly last us beyond three days, within which time it was not in any degree probable that we should find relief from our present sufferings, and that hunger would at last drive us to the necessity of casting lots. It was without any objection agreed to, and we set to work as fast as we were able to prepare it so as to prevent its spoiling. We separated his limbs from his body, and cut all the flesh from the bones; after which, we opened the body, took out the heart, and then closed it again—sewed it up as decently as we could, and committed it to the sea. We now first commenced to satisfy the immediately craving of nature from the heart, which we eagerly devoured, and then eat sparingly of a few pieces of the flesh; after which, we hung up the remainder, cut in thin strips about the boat, to dry in the sun: we made a fire and roasted some of it, to serve us during the next day. In this manner did we dispose of our fellow-sufferer; the painful recollection of which, brings to mind at this moment, some of the most disagreeable and revolting ideas that it is capable of conceiving. We knew not then, to whose lot it would fall next, either to die or be shot, and eaten like the poor wretch we had just dispatched. Humanity must shudder at the dreadful recital. I have no language to paint the anguish of our souls in this dreadful dilemma. The next morning, the 10th of February, we found that the flesh had become tainted, and had turned of a greenish colour, upon which we concluded to make a fire and cook it at once, to prevent its becoming so putrid as not to be eaten at all: we accordingly did so, and by that means preserved it for six or seven days longer; our bread during the time, remained untouched; as that would not be liable to spoil, we placed it carefully aside for the last moments of our trial. About three o'clock this afternoon a strong breeze set in from the NW. and we made very good progress, considering that we were compelled to steer the boat by management of the sails alone: this wind continued until the thirteenth, when it changed again ahead. We contrived to keep soul and body together by sparingly partaking of our flesh, cut up in small pieces and eaten with salt-water. By the fourteenth, our bodies became so far recruited, as to enable us to make a few attempts at guiding our boat again with the oar; by each taking his turn, we managed to effect it, and to make a tolerable good course. On the fifteenth, our flesh was all consumed, and we were driven to the last morsel of bread, consisting of two cakes; our limbs had for the last two days swelled very much, and now began to pain us most excessively. We were still, as near as we could judge, three hundred miles from the land, and but three days of our allowance on hand. The hope of a continuation of the wind, which came out at west this morning, was the only comfort and solace that remained to us: so strong had our desires at last reached in this respect, that a high fever had set in, in our veins, and a longing that nothing but its continuation could satisfy. Matters were now with us at their height; all hope was cast upon the breeze; and we tremblingly and fearfully awaited its progress, and the dreadful development of our destiny. On the sixteenth, at night, full of the horrible reflections of our situation, and panting with weakness, I laid down to sleep, almost indifferent whether I should ever see the light again. I had not lain long, before I dreamt I saw a ship at some distance off from us, and strained every nerve to get to her, but could not. I awoke almost overpowered with the frenzy I had caught in my slumbers, and stung with the cruelties of a diseased and disappointed imagination. On the seventeenth, in the afternoon, a heavy cloud appeared to be settling down in an E. by N. direction from us, which in my view, indicated the vicinity of some land, which I took for the island of

Massafuera. I concluded it could be no other; and immediately upon this reflection, the life blood began to flow again briskly in my veins. I told my companions that I was well convinced it was land, and if so, in all probability we would reach it before two days more. My words appeared to comfort them much; and by repeated assurances of the favourable appearance of things, their spirits acquired even a degree of elasticity that was truly astonishing. The dark features of our distress began now to diminish a little, and the countenance, even amid the gloomy bodings of our hard lot, to assume a much fresher hue. We directed our course for the cloud, and our progress that night was extremely good. The next morning, before daylight, Thomas Nicholson, a boy about seventeen years of age, one of my two companions who had thus far survived with me, after having bailed the boat, laid down, drew a piece of canvass over him, and cried out, that he then wished to die immediately. I saw that he had given up, and I attempted to speak a few words of comfort and encouragement to him, and endeavoured to persuade him that it was a great weakness and even wickedness to abandon a reliance upon the Almighty, while the least hope, and a breath of life remained; but he felt unwilling to listen to any of the consolatory suggestions which I made to him; and, notwithstanding the extreme probability, which I stated there was of our gaining the land before the end of two days more, he insisted upon lying down and giving himself up to despair. A fixed look of settled and forsaken despondency came over his face: he lay for some time silent, sullen, and sorrowful—and I felt at once satisfied, that the coldness of death was fast gathering upon him: there was a sudden and unaccountable earnestness in his manner, that alarmed me, and made me fear that I myself might unexpectedly be overtaken by a like weakness, or dizziness of nature, that would bereave me at once of both reason and life, but Providence willed it otherwise.

At about seven o'clock this morning, while I was lying asleep, my companion who was steering, suddenly and loudly called out, "There's a Sail!" I know not what was the first movement I made upon hearing such an unexpected cry: the earliest of my recollections are, that immediately I stood up, gazing in a state of abstraction and ecstasy upon the blessed vision of a vessel about seven miles off from us; she was standing in the same direction with us, and the only sensation I felt at the moment was, that of a violent and unaccountable impulse to fly directly towards her. I do not believe it is possible to form a just conception of the pure, strong feelings, and the unmingled emotions of joy and gratitude, that took possession of my mind on this occasion: the boy, too, took a sudden and animated start from his despondency, and stood up to witness the probable instrument of his salvation. Our only fear was now, that she would not discover us, or that we might not be able to intercept her course: we, however, put our boat immediately, as well as we were able, in a direction to cut her off; and found, to our great joy, that we sailed faster than she did. Upon observing us, she shortened sail, and allowed us to come up to her. The captain hailed us and asked who we were. I told him we were from a wreck, and he cried out immediately for us to come alongside the ship. I made an effort to assist myself along to the side, for the purpose of getting up, but strength failed me altogether, and I found it impossible to move a step further without help. We must have formed at that moment, in the eyes of the captain and his crew, a most deplorable and affecting picture of suffering and misery. Our cadaverous countenances, sunken eyes, and bones just starting through our skin, with the ragged remnants of clothes stuck about our sun-burnt bodies, must have produced an appearance to him affecting and revolting in the highest degree. The sailors commenced to remove us from our boat, and we were taken to the cabin, and comfortably provided for in every respect. In a few minutes we were permitted to taste of a little thin food, made from tapiocha, and in a few days with prudent management, we were considerably recruited. This vessel proved to be the brig Indian, captain William Crozier, of London; to whom we are indebted for every polite, friendly, and attentive disposition towards us, that can possibly characterize a man of humanity and feeling. We were taken up in latitude 33° 45' S. longitude 81° 03' W. At twelve o'clock this day we saw the island of Massafuera, and on the 25th of February, we arrived at Valparaiso in utter distress and poverty. Our wants were promptly relieved there.

The captain and the survivors of his boat's crew, were taken up by the American whale-ship, the Dauphin, Captain Zimri Coffin, of Nantucket, and arrived at Valparaiso on the seventeenth of March following: he was taken up in latitude 37°; S. off the island of St. Mary. The third boat got separated from him on the 28th of January, and has not been heard of since. The names of all the survivors, are as follows: Captain George Pollard, junr. Charles Ramsdale, Owen Chase, Benjamin Lawrence, and Thomas Nicholson, all of Nantucket. There died in the captain's boat, the following: Brazilla Ray of Nantucket, Owen Coffin of the same place, who was shot, and Samuel Reed, a black.

The captain relates, that after being separated, as herein before stated, they continued to make what progress they could towards the island of Juan Fernandez, as was agreed upon; but contrary winds and the extreme debility of the crew prevailed against their united exertions. He was with us equally surprised and concerned at the separation that took place between us; but continued on his course, almost confident of meeting with us again. On the fourteenth, the whole stock of provisions belonging to the second mate's boat, was entirely exhausted, and on the twenty-fifth, the black man, Lawsen Thomas, died, and was eaten by his surviving companions. On the twenty-first, the captain and his crew were in the like dreadful situation with respect to their provisions; and on the twenty-third, another coloured man, Charles Shorter, died out of the same boat, and his body was shared for food between the crews of both boats. On the twenty-seventh, another, Isaac Shepherd (a black man), died in the third boat; and on the twenty-eight, another black, named Samuel Reed, died out of the captain's boat. The bodies of these men constituted their only food while it lasted; and on the twenty-ninth, owing to the darkness of the night and want of sufficient power to manage their boats, those of the captain and second mate separated in latitude 35°; S. longitude 100°; W. on the 1st of February, having consumed the last morsel, the captain and the three other men that remained with him, were reduced to the necessity of casting lots. It fell upon Owen Coffin to die, who with great fortitude and resignation submitted to his fate. They drew lots to see who should shoot him: he placed himself firmly to receive his death, and was immediately shot by Charles Ramsdale, whose hard fortune it was to become his executioner. On the 11th, Brazilla Ray died; and on these two bodies the captain and Charles Ramsdale, the only two that were then left, subsisted until the morning of the twenty-third, when they fell in with the ship Dauphin, as before stated, and were snatched from impending destruction. Every assistance and attentive humanity, was bestowed upon them by Capt. Coffin, to whom Capt. Pollard acknowledged every grateful obligation. Upon making known the fact, that three of our companions had been left at Ducies Island, to the captain of the U.S. frigate Constellation, which lay at Valparaiso when we arrived, he said he should immediately take measures to have them taken off.

On the 11th of June following I arrived at Nantucket in the whale-ship the Eagle, Capt. William H. Coffin. My family had received the most distressing account of our shipwreck, and had given me up for lost. My unexpected appearance was welcomed with the most grateful obligations and acknowledgments to a beneficent Creator, who had guided me through darkness, trouble, and death, once more to the bosom of my country and friends.

Supplement

The following is a list of the whole crew of the ship, with their arrangements into the three several boats upon starting from the wreck: the names of those who died, were left on the island, or shot—with those also who survived, and who were in the third or second mate's boat at the time of separation—and whose fate is yet uncertain:

Capt. James Pollard, jun., 1st boat, survived
Obed Hendricks, 1st boat, put in 3rd boat
Brazilla Ray, 1st boat, died
Owen Coffin, 1st boat, shot
Samuel Reed (black), 1st boat, died
Charles Ramsdale, 1st boat, survived
Seth Weeks, 1st boat, left on the island
Owen Chase, 2nd boat, survived
Benjamin Lawrence, 2nd boat, survived
Thomas Nicholson, 2nd boat, survived
Isaac Cole, 2nd boat, died
Richard Peterson (black), 2nd boat, died
William Wright, 2nd boat, left on the island
Matthew P. Joy, 3rd boat, died
Thomas Chapple, 3rd boat, left on the island
Joseph West, 3rd boat, missing
Lawson Thomas (black), 3rd boat, died
Charles Shorter (black), 3rd boat, died
Isaiah Shepherd (black), 3rd boat, died
William Bond (black), 3rd boat, missing

Finis

JEREMIAH N. REYNOLDS, "MOCHA DICK; OR THE WHITE WHALE OF THE PACIFIC: A LEAF FROM A MANUSCRIPT JOURNAL"

KNICKERBOCKER
12 (MAY 1839), 377–392

INTRODUCTION

In 1847, Melville purchased Reynolds's (1799–1858) *Voyage of the United States Frigate Potomac* (1835), and most scholars believe that he must have also been familiar with "Mocha Dick," clearly the best candidate for one of the "wild legends" Melville wrote of to his English publisher Richard Bentley in the spring of 1850. Reynolds, a persistent lobbyist for a national exploring expedition to the Pacific, was part of the circle of veteran sealers and entrepreneurs which included Nathaniel B. Palmer (1799–1877), reputed discoverer of Antarctica and namesake of the captain named below in Reynolds's story.

If Owen Chase's book be claimed as the inspiration of *Moby-Dick*—at least for the climactic sinking of the *Pequod*—then Reynolds's tale has an equal claim to inspiring the greatest animal character in fiction. Readers should note not only the description of the white whale itself, but also the implicit comparison of switchel (ginger-jub in Chapter 72 of *Moby-Dick*) and rum—a detail that suggests that Melville revisited Reynolds and was not simply recalling a general impression. John Huston and Ray Bradbury also seem to have drawn on Reynolds for the final resolve they give to Starbuck at the conclusion of the 1956 film version of *Moby Dick*: see Jon Eller, *Ray Bradbury Unbound* (Indianapolis: Indiana University Press, 2015). Bercaw #582. Modern edition: in Parker and Hayford (2002).

PRIMARY SOURCE

We expected to find the island of Santa Maria still more remarkable for the luxuriance of its vegetation, than even the fertile soil of Mocha; and the disappointment arising from the unexpected shortness of our stay at the latter place, was in some degree relieved, by the prospect of our remaining for several days in safe anchorage at the former. Mocha lies upon the coast of Chili, in lat. 38° 28' south, twenty leagues north of Mono del Bonifacio, and opposite the Imperial river, from which it bears w. s. w. During the last century, this island was inhabited by the Spaniards, but it is at present, and has been for some years, entirely deserted. Its climate is mild, with little perceptible difference of temperature between the summer and winter seasons. Frost is unknown on the lowlands, and snow is rarely seen, even on the summits of the loftiest mountains.

It was late in the afternoon, when we left the schooner; and while we bore up for the north, she stood away for the southern extremity of the island. As evening was gathering around us, we fell in with a vessel, which proved to be the same whose boats, a day or two before, we had seen in the act of taking a whale. Aside from the romantic and stirring associations it awakened, there are few objects in themselves more picturesque or beautiful, than a whale-ship, seen from a distance of three or four miles, on a pleasant evening, in the midst of the great Pacific. As she moves gracefully over the water, rising and falling on the gentle undulations peculiar to this sea; her sails glowing in the quivering light of the fires that flash from below, and a thick volume of smoke ascending from the midst, and curling away in dark masses upon the wind; it requires little effort of the fancy, to imagine one's self gazing upon a floating volcano.

As we were both standing to the north, under easy sail, at nine o'clock at night we had joined company with the stranger. Soon after, we were boarded by his whale-boat, the officer in command of which bore us the compliments of the captain, together with a friendly invitation to partake the hospitalities of his cabin. Accepting, without hesitation, a courtesy so frankly tendered, we proceeded, in company with Captain Palmer, on board, attended by the mate of the *Penguin,* who was on his way to St Mary's to repair his boat, which had some weeks before been materially injured in a storm.

We found the whaler a large, well-appointed ship, owned in New-York, and commanded by such a man as one might expect to find in charge of a vessel of this character; plain, unassuming, intelligent, and well-informed upon all the subjects relating to his peculiar calling. But what shall we say of his first mate, or how describe him? To attempt his portrait by a comparison, would be vain, for we have never looked upon his like, and a detailed description, however accurate, would but faintly shadow forth the *tout ensemble* of his extraordinary figure. He had probably numbered about thirty-five years. We arrived at this conclusion, however, rather from the untamed brightness of his flashing eye, than the general appearance of his features, on which torrid sun and polar storm had left at once the furrows of more advanced age, and a tint swarthy as that of the Indian. His height, which was a little beneath the common standard, appeared almost dwarfish, from the immense breadth of his overhanging shoulders, while the unnatural length of the loose, dangling arms which hung from them, and which, when at rest, had least the appearance of ease, imparted to his uncouth and muscular frame an air of grotesque awkwardness, which defies description. He made few pretensions as a sailor, and had never aspired to the command of a ship. But he would not have exchanged the sensations which stirred his blood, when steering down upon a school of whales, for the privilege of treading, as master, the deck of the noblest liner that ever traversed the Atlantic. According to the admeasurement of his philosophy, whaling was the most dignified and manly of all sublunary pursuits. Of this he felt perfectly satisfied, having been engaged in the noble vocation for upward of twenty years, during which period, if his own assertions were to be received as evidence, no man in the American spermaceti fleet had made so many captures, or met with such wild adventures, in the exercise of his perilous profession. Indeed, so completely were all his propensities, thoughts, and feelings, identified with his occupation; so intimately did he seem acquainted with the habits and instincts of the objects of his pursuit, and so little conversant with the ordinary affairs of life; that one felt less inclined to class him in the genus *homo,* than as a sort of intermediate something between man and the cetaceous tribe.

Soon after the commencement of his nautical career, in order to prove that he was not afraid of a whale, a point which it is essential for the young whaleman to establish beyond question, he offered, upon a wager, to run his boat "bows on" against the side of an "old bull", leap from the "cuddy" to the back of the fish, sheet his lance home, and return on board in safety. This feat, daring as it may be considered, he undertook and accomplished; at least so it was chronicled in his log, and he was ready to bear witness, on oath, to the veracity of the record. But his conquest of the redoubtable MOCHA DICK, unquestionably formed the climax of his exploits.

Before we enter into the particulars of this triumph, which, through their valorous representative, conferred so much honor on the lancers of Nantucket, it may be proper to inform the reader who and what Mocha Dick was; and thus give him a posthumous introduction to one who was, in his day and generation, so emphatically among fish the "Stout Gentleman" of his latitudes. The introductory portion of his history we shall give, in a condensed form, from the relation of the mate. Substantially, however, it will be even as he rendered it; and as his subsequent narrative, though not deficient in rude eloquence, was coarse in style and language, as well as unnecessarily diffuse, we shall assume the liberty of altering the expression; of adapting the phraseology to the occasion; and of presenting the whole matter in a shape more succinct and connected. In this arrangement, however, we shall leave our adventurer to tell his *own story,* although not always in his own words, and shall preserve the person of the original.

But to return to Mocha Dick—which, it may be observed, few were solicitous to do, who had once escaped from him. This renowned monster, who had come off victorious in a hundred fights with his pursuers, was an old bull whale, of prodigious size and strength. From the effect of age, or more probably from a freak of nature, as exhibited in the case of the Ethiopian Albino, a singular consequence had resulted—*he was white as wool!* Instead of projecting his spout obliquely forward, and puffing with a short, convulsive effort, accompanied by a snorting noise, as usual with his species, he flung the water from his nose in a lofty, perpendicular, expanded volume, at regular and somewhat distant intervals; its expulsion producing a continuous roar, like that of vapor struggling from the safety-valve of a powerful steam engine. Viewed from a distance, the practised eye of the sailor only could decide, that the moving mass, which constituted this enormous animal, was not a white cloud sailing along the horizon. On the spermaceti whale, barnacles are rarely discovered; but upon the head of this *lusus naturae,* they had clustered, until it became absolutely rugged with the shells. In short, regard him as you would, he was a most extraordinary fish; or, in the vernacular of Nantucket, "a genuine old sog", of the first water.

Opinions differ as to the time of his discovery. It is settled, however, that previous to the year 1810, he had been seen and attacked near the island of Mocha. Numerous boats are known to have been shattered by his immense flukes, or ground to pieces in the crush of his powerful jaws; and, on one occasion, it is said that he came off victorious from a conflict with the crews of three English whalers, striking fiercely at the last of the retreating boats, at the moment it was rising from the water, in its hoist up to the ship's davits. It must not be supposed, howbeit, that through all this desperate warfare, our leviathan passed scathless. A back serried with irons, and from fifty to a hundred yards of line trailing in his wake, sufficiently attested, that though unconquered, he had not proved invulnerable. From the period of Dick's first appearance, his celebrity continued to increase, until his name seemed naturally to mingle with the salutations which whalemen were in the habit of exchanging, in their encounters upon the broad Pacific; the customary interrogatories almost always closing with, "Any news from Mocha Dick?" Indeed, nearly every whaling captain who rounded Cape Horn, if he possessed any professional ambition, or valued himself on his skill in subduing the monarch of the seas, would lay his vessel along the coast, in the hope of having an opportunity to try the muscle of this doughty champion, who was never known to shun his assailants. It was remarked, nevertheless, that the old fellow seemed particularly careful as to the portion of his body which he exposed to the approach of the boat-steerer; generally presenting, by some well-timed manoeuvre, his back to the harpooneer; and dexterously evading every attempt to plant an iron under his fin, or a spade on his "small". Though naturally fierce, it was not customary with Dick, while unmolested, to betray a malicious disposition. On the contrary, he would sometimes pass quietly round a vessel, and occasionally swim lazily and harmlessly among the boats, when armed with full craft, for the destruction of his race. But this forbearance gained him little credit, for if no other cause of accusation remained to them, his foes would swear they saw a lurking deviltry in the long, careless sweep of his flukes. Be this as it may, nothing is more certain, than that all indifference vanished with the first prick of the harpoon; while cutting the line, and a hasty retreat to their vessel, were frequently the only means of escape from destruction, left to his discomfited assaulters.

Thus far the whaleman had proceeded in his story and was about commencing the relation of his own individual encounters with its subject, when he was cut short by the mate of the *Penguin*, to whom allusion has already been made, and who had remained, up to this point, an excited and attentive listener. Thus he would have continued, doubtless, to the end of the chapter, notwithstanding his avowed contempt for every other occupation than sealing, had not an observation escaped the narrator, which tended to arouse his professional jealousy. The obnoxious expression we have forgotten. Probably it involved something of boasting or egotism; for no sooner was it uttered, than our sealer sprang from his seat, and planting himself in front of the unconscious author of the insult, exclaimed:

"*You!*—you whale-killing, blubber-hunting, light-gathering varmint!—you pretend to manage a boat better than a Stonington sealer! A Nantucket whaleman," he continued, curling his lip with a smile of supreme disdain, "presume to teach a Stonington sealer how to manage a boat! Let all the small craft of your South Sea fleet range among the rocks and breakers where I have been, and if the whales would not have a peaceful time of it, for the next few years, may I never strip another jacket, or book another skin! What's taking a whale? Why, I could reeve a line through one's blow-hole, make it fast to a thwart, and then beat his brains out with my seal-club!"

Having thus given play to the first ebullition of his choler, he proceeded with more calmness to institute a comparison between whaling and sealing. "A whaler," said he, "never approaches land, save when he enters a port to seek fresh grub. Not so the sealer. *He* thinks that his best fortune, which leads him where the form of man has never before startled the game he's after; where a quick eye, steady nerve, and stout heart, are his only guide and defence, in difficulty and danger. Where the sea is roughest, the whirlpool wildest, and the surf roars and dashes madly among the jagged cliffs, there—I was going to say there *only*—are the peak-nosed, black-eyed rogues we hunt for, to be found, gambolling in the white foam, and there must the sealer follow them. Were I to give you an account of my adventures about the Falkland Isles; off the East Keys of Staten Land; through the South Shetlands; off the Cape, where we lived on salt pork and seal's flippers; and finally, the story of a season spent with a single boat's crew on Diego Ramirez,[1] you would not make such a fuss about your Mocha Dick. As to the straits of Magellen, Sir, they are as familiar to me, as Broadway to a New-York dandy; though *it* should strut along that fashionable promenade twelve dozen times a day."

Our son of the sea would have gone on to particularize his "hair-breadth 'scapes and moving accidents", had we not interposed, and insisted that the remainder of the night should be devoted to the conclusion of Dick's history; at the same time assuring the "knight of the club" that so soon as we met at Santa Maria, he should have an entire evening expressly set apart, on which he might glorify himself and his calling. To this he assented, with the qualification, that his compliance with the general wish, in thus yielding precedence to his rival, should not be construed into an admission, that Nantucket whalemen were the best boatmen in the world, or that sealing was not as honorable and as pretty a business for coining a penny, as the profession of "blubber-hunting" ever was.

The whaler now resumed. "I will not weary you," said he, "with the uninteresting particulars of a voyage to Cape Horn. Our vessel, as capital a ship as ever left the little island of Nantucket, was finely manned and commanded, as well as thoroughly provided with every requisite for the peculiar service in which she was engaged. I may here observe, for the information of such among you as are not familiar with these things, that soon after a whale-ship from the United States is fairly at sea, the men are summoned aft; then boats' crews are selected by the captain and first mate, and a ship-keeper, at the same time, is usually chosen. The place to be filled by this individual is an important one and the person designated should be a careful and sagacious man. His duty is, more particularly, to superintend the vessel while the boats are away, in chase of fish; and at these times, the cook and steward are perhaps his only crew. His station, on these occasions, is at the mast-head, except when he is wanted below to assist in working the ship. While aloft, he is to look out for whales, and also to keep a strict and tireless eye upon the absentees, in order to render them immediate assistance, should emergency require it. Should the game rise to windward of their pursuers, and they be too distant to observe personal signs, he must run down the jib. If they rise to leeward he should haul up the spanker; continuing the little black signal-flag at the mast; so long as they remain on the surface. When the "school" turn flukes, and go down, the flag is to be struck, and again displayed when they are seen to ascend. When circumstances occur which require the return of the captain on board, the colors are to be hoisted at the mizzen peak. A ship-keeper must farther be sure that provisions are ready for the men, on their return from the chase, and that drink be amply furnished, in the form of a bucket of "switchel". "No whale, no switchel", is frequently the rule; but I am inclined to think that, whale or no whale, a little rum is not amiss, after a lusty pull.

"I have already said, that little of interest occurred, until after we had doubled Cape Horn. We were now standing in upon the coast of Chili, before a gentle breeze from the south, that bore us along almost imperceptibly. It was a quiet and beautiful evening, and the sea glanced and glistened in the level rays of the descending sun, with a surface of waving gold. The western sky was flooded with amber light, in the midst of which, like so many islands, floated immense clouds, of every conceivable brilliant dye; while far to the northeast, looming darkly against a paler heaven, rose the conical peak of Mocha. The men were busily employed in sharpening their harpoons, spades, and lances, for the expected fight. The look-out at the mast-head, with cheek on his shoulder, was dreaming of the 'dangers he had passed', instead of keeping watch for those which were to come; while the captain paced the quarter-deck with long and hasty stride, scanning the ocean in every direction, with a keen, expectant eye. All at once, he stopped, fixed his gaze intently for an instant on some object to leeward, that seemed to attract it, and then, in no very conciliating tone, hailed the mast-head:

"'Both ports shut?' he exclaimed, looking aloft, and pointing backward, where a long white bushy spout was rising, about a mile off the larboard bow, against the glowing horizon. 'Both ports shut? I say, you leaden-eyed lubber! Nice lazy son of a sea-cook *you* are, for a look-out! Come down, Sir!'

"'There she blows!—sperm whale—old sog, sir;' said the man, in a deprecatory tone, as he descended from his nest in the air. It was at once seen that the creature was companionless; but as a lone whale is generally an old bull, and of unusual size and ferocity, more than ordinary sport was anticipated, while unquestionably more than ordinary honor was to be won from its successful issue.

"The second mate and I were ordered to make ready for pursuit; and now commenced a scene of emulation and excitement, of which the most vivid description would convey but an imperfect outline, unless you have been a spectator or an actor on a similar occasion. Line-tubs, water-kegs, and waif-poles, were thrown hurriedly into the boats; the irons were placed in the racks, and the necessary evolutions of the ship gone through, with a quickness almost magical; and this too, amidst what to a landsman would have seemed inextricable confusion, with perfect regularity and precision; the commands of the officers being all but forestalled by the enthusiastic eagerness of the men. In a short time, we were as near the object of our chase, as it was considered prudent to approach.

"'Back the main-top-s'l!' shouted the captain. 'There she blows! there she blows!—there she blows!'—cried the look-out, who had taken the place of his sleepy shipmate, raising the pitch of his voice with each announcement, until it amounted to a downright yell: 'Right ahead, Sir!—spout as long an 's thick as the mainyard!'

"'Stand by to lower!' exclaimed the captain; 'all hands; cook, steward, cooper—every d—d one of ye, stand by to lower!'

"An instantaneous rush from all quarters of the vessel answered this appeal, and every man was at his station, almost before the last word had passed the lips of the skipper.

"'Lower away!'—and in a moment the keels splashed in the water. 'Follow down the crews; jump in my boys; ship the crotch; line your oars; now pull, as if the d—l was in your wake!' were the successive orders, as the men slipped down the ship's side, took their places in the boats, and began to give way.

"The second mate had a little the advantage of me in starting. The stern of his boat grated against the bows of mine, at the instant I grasped my steering-oar, and gave the word to shove off. One sweep of my arm, and we sprang foaming in his track. Now came the tug of war. To become a first-rate oarsman, you must understand, requires a natural gift. My crew were not wanting in the proper qualification; every mother's son of them pulled as if he had been born with an oar in his hand; and as they stretched every sinew for the glory of darting the first iron it did my heart good to see the boys spring. At every stroke, the tough blades bent like willow wands, and quivered like tempered steel in the warm sunlight, as they sprang forward from the tension of the retreating wave. At the distance of half a mile and directly before us, lay the object of our emulation and ambition, heaving his huge bulk in unwieldy gambols, as though totally unconscious of our approach.

"'There he blows! An old bull, by Jupiter! Eighty barrels, boys, waiting to be towed alongside! Long and quick—shoot ahead! Now she feels it; waist boat never could beat us, now she feels the touch!—now she walks through it! Again—*now!*' Such were the broken exclamations and adjurations with which I cheered my rowers to their toil, as, with renewed vigor, I plied my long steering-oar. In another moment, we were alongside our competitor. The shivering blades flashed forward and backward, like sparks of light. The waters boiled under our prow, and the trenched waves closed, hissing and whirling, in our wake, as we swept, I might almost say were *lifted,* onward in our arrowy course.

"We were coming down upon our fish, and could hear the roar of his spouting above the rush of the sea, when my boat began to take the lead.

"'Now, my fine fellows,' I exclaimed, in triumph, 'now we'll show them our stern—only spring! Stand ready, harpooner, but don't dart, till I give the word.'

"'Carry me on, and his name's *Dennis!*'[2] cried the boat-steerer, in a confident tone. We were perhaps a hundred feet in advance of the waist-boat, and within fifty of the whale, about an inch of whose hump only was to be seen above the water, when, heaving slowly into view a pair of flukes some eighteen feet in width, he went down. The men lay on their oars. 'There he blows, again!' cried the tub-oarsman, as a lofty, perpendicular spout sprang into the air, a few furlongs away on the starboard side. Presuming from his previous movement, that the old fellow had been 'gallied' by other boats, and might probably be jealous of our purpose, I was about ordering the men to pull away as softly and silently as possible, when we received fearful intimation that he had no intention of balking our inclination, or even yielding us the honor of the first attack. Lashing the sea with his enormous tail, until he threw about him a cloud of surf and spray, he came down, at full speed, 'jaws on', with the determination, apparently, of doing battle in earnest. As he drew near, with his long curved back looming occasionally above the surface of the billows, we perceived that it was *white as the surf around him;* and the men stared aghast at each other, as they uttered, in a suppressed tone, the terrible name of MOCHA DICK!

"'Mocha Dick or the d—l,' said I, 'this boat never sheers off from any thing that wears the shape of a whale. Pull easy; just give her way enough to steer.' As the creature approached, he somewhat abated his frenzied speed, and, at the distance of a cable's length, changed his course to a sharp angle with our own.

"'Here he comes!' I exclaimed. 'Stand up, harpooner! Don't be hasty—don't be flurried. Hold your iron higher—firmer. Now!' I shouted, as I brought our bows within a boat's length of the immense mass which was wallowing heavily by. '*Now!—give it to him solid!*'

"But the leviathan plunged on, unharmed. The young harpooner, though ordinarily as fearless as a lion, had imbibed a sort of superstitious dread of Mocha Dick, from the exaggerated stories of that prodigy, which he had heard from his comrades. He regarded him, as he had heard him described in many a tough yarn during the middle watch, rather as some ferocious fiend of the deep, than a regular-built, legitimate whale! Judge then of his trepidation, on beholding a creature, answering the wildest dreams of his fancy, and sufficiently formidable, without any superadded terrors, bearing down upon him with thrashing flukes and distended jaws! He stood erect, it cannot be denied. He planted his foot—he grasped the coil—he poised his weapon. But his knee shook, and his sinewy arm wavered. The shaft was hurled, but with unsteady aim. It just grazed the back of the monster, glanced off, and darted into the sea beyond. A second, still more abortive, fell short of the mark. The giant animal swept on for a few rods, and then, as if in contempt of our fruitless and childish attempts to injure him, flapped a storm of spray in our faces with his broad tail, and dashed far down into the depths of the ocean, leaving our little skiff among the waters where he sank, to spin and duck in the whirlpool.

"Never shall I forget the choking sensation of disappointment which came over me at that moment. My glance fell on the harpooner. 'Clumsy lubber!' I vociferated, in a voice hoarse with passion; '*you* a whaleman! You are only fit to spear eels! Cowardly spawn! Curse me, if you are not *afraid* of a whale!'

"The poor fellow, mortified at his failure, was slowly and thoughtfully hauling in his irons. No sooner had he heard me stigmatize him as 'afraid of a whale', than he bounded upon his thwart, as if bitten by a serpent. He stood before me for a moment, with a glowing cheek and flashing eye; then, dropping the iron he had just drawn in, without uttering a word, he turned half round, and sprang head-foremost into the sea. The tub-oarsman, who was recoiling the line in the after part of the boat, saw his design just in season to grasp him by the heel, as he made his spring. But he was not to be dragged on board again without a struggle. Having now become more calm, I endeavored to soothe his wounded pride with kind and flattering words; for I knew him to be a noble-hearted fellow, and was truly sorry that my hasty reproaches should have touched so fine a spirit so deeply.

"Night being now at hand, the captain's signal was set for our return to the vessel; and we were soon assembled on her deck, discussing the mischances of the day, and speculating on the prospect of better luck on the morrow.

"We were at breakfast next morning, when the watch at the fore-top-gallant head sung out merrily, 'There she breaches!' In an instant every one was on his feet. 'Where away?' cried the skipper, rushing from the cabin, and upsetting in his course the steward, who was returning from the caboose with a replenished biggin of hot coffee. 'Not loud but deep' were the grumblings and groans of that functionary, as he rubbed his scalded shins, and danced about in agony; but had they been far louder, they would have been drowned in the tumult of vociferation which answered the announcement from the mast-head.

"'Where away?' repeated the captain, as he gained the deck.

"'Three points off the leeward bow.'

"'How far?'

"'About a league, Sir; heads same as we do. There she blows!' added the man, as he came slowly down the shrouds, with his eyes fixed intently upon the spouting herd.

"'Keep her off two points! Steady!—steady, as she goes!'

"'Steady it is, Sir,' answered the helmsman.

"'Weather braces, a small pull. Loose to'-gallant-s'ls! Bear a hand, my boys! Who knows but we may tickle their ribs at this rising?'

"The captain had gone aloft, and was giving these orders from the main-to'-gallant-cross-trees. 'There she top-tails! there she blows!' added he, as, after taking a long look at the sporting shoal, he glided down the back stay. 'Sperm whale, and a thundering big school of 'em!' was his reply to the rapid and eager inquiries of the men. 'See the lines in the boats,' he continued; 'get in the craft; swing the cranes!'

"By this time the fish had gone down, and every eye was strained to catch the first intimation of their reappearance. "'There she *spouts!*' screamed a young greenhorn in the main chains, 'close by; a mighty big whale, Sir!'

"'We'll know that better at the trying out, my son,' said the third mate, drily.

"'Back the main-top-s'l!' was now the command. The ship had little headway at the time, and in a few minutes we were as motionless as if lying at anchor.

"'Lower away, all hands!' And in a twinkling, and together, the starboard, larboard, and waist-boats struck the water. Each officer leaped into his own; the crews arranged themselves at their respective stations; the boat-steerers began to adjust their 'craft'; and we left the ship's side in company; the captain, in laconic phrase, bidding us to 'get up and get fast', as quickly as possible.

"Away we dashed, in the direction of our prey, who were frolicking, if such a term can be applied to their unwieldy motions, on the surface of the waves. Occasionally, a huge, shapeless body would flounce out of its proper element, and fall back with a heavy splash; the effort forming about as ludicrous a caricature of agility, as would the attempt of some over-fed alderman to execute the Highland fling.

"We were within a hundred rods of the herd, when, as if from a common impulse, or upon some preconcerted signal, they all suddenly disappeared. 'Follow me!' I shouted, waving my hand to the men in the other boats; 'I see their track under water; they swim fast, but we'll be among them when they rise. Lay back,' I continued, addressing myself to my own crew, 'back to the thwarts! Spring *hard!* We'll be in the thick of 'em when they come up; only *pull!*'

"And they did pull, manfully. After towing for about a mile, I ordered them to 'lie'. The oars were peaked, and we rose to look out for the first 'noddle-head' that should break water. It was at this time a dead calm. Not a single cloud was passing over the deep blue of the heavens, to vary their boundless transparency, or shadow for a moment the gleaming ocean which they spanned. Within a short distance lay our noble ship, with her idle canvass hanging in drooping festoons from her yards; while she seemed resting on her inverted image, which, distinct and beautiful as its original, was glassed in the smooth expanse beneath. No sound disturbed the general silence, save our own heavy breathings, the low gurgle of the water against the side of the boat, or the noise of flapping wings, as the albatross wheeled sleepily along through the stagnant atmosphere. We had remained quiet for about five minutes, when some dark object was descried ahead, moving on the surface of the sea. It proved to be a small 'calf', playing in the sunshine.

"'Pull up and strike it,' said I to the third mate; 'it may bring up the old one—perhaps the whole school.'

"And so it did, with a vengeance! The sucker was transpierced, after a short pursuit; but hardly had it made its first agonized plunge, when an enormous cow-whale rose close beside her wounded offspring. Her first endeavor was to take it under her fin, in order to bear it away; and nothing could be more striking than the maternal tenderness she manifested in her exertions to accomplish this object. But the poor thing was dying, and while she vainly tried to induce it to accompany her, it rolled over, and floated dead at her side. Perceiving it to be beyond the reach of her caresses, she turned to wreak her vengeance on its slayers, and made directly for the boat, crashing her vast jaws the while, in a paroxysm of rage. Ordering his boat-steerer aft, the mate sprang forward, cut the line loose from the calf, and then snatched from the crotch the remaining iron, which he plunged with his gathered strength into the body of the mother, as the boat sheered off to avoid her onset. I saw that the work was well done, but had no time to mark the issue; for at that instant, a whale 'breached' at the distance of about a mile from us, on the starboard quarter. The glimpse I caught of the animal in his descent, convinced me that I once more beheld my old acquaintance, Mocha Dick. That falling mass was white as a snow-drift!

"One might have supposed the recognition mutual, for no sooner was his vast square head lifted from the sea, than he charged down upon us, scattering the billows into spray as he advanced, and leaving a wake of foam a rod in width, from the violent lashing of his flukes.

"'He's making for the bloody water!' cried the men, as he cleft [*sic*] his way toward the very spot where the calf had been killed. 'Here, harpooner, steer the boat, and let me dart!' I exclaimed, as I leaped into the bows. 'May the "*Goneys*" eat me, if he dodge us *this* time. though he were Beelzebub himself! Pull for the red water!'

"As I spoke, the fury of the animal seemed suddenly to die away. He paused in his career, and lay passive on the waves; with his arching back thrown up like the ridge of a mountain. 'The old sog's lying to!' I cried, exultingly. 'Spring, boys! spring *now,* and we have him! All my clothes, tobacco, every thing I've got, shall be yours, only lay me 'longside that whale before another boat comes up! My *grimky!* what a hump! Only look at the irons in his back! No, don't *look*—PULL! Now, boys, if you care about seeing your sweethearts and wives in old Nantuck!—if you love Yankee-land—if you love *me*—pull ahead, *won't* ye? Now then, to the thwarts! Lay back, my boys! I feel ye, my hearties! Give her the touch. Only five seas off! *Not* five seas off! One minute—*half* a minute more! Softly—no noise. Softly with your oars! That will do—'

"And as the words were uttered, I raised the harpoon above my head, took a rapid but no less certain aim, and sent it, hissing, deep into his thick white side!

"'Stern all! for your lives!' I shouted; for at that instant the steel quivered in his body, the wounded leviathan plunged his head beneath the surface, and whirling around with great velocity, smote the sea violently, with fin and fluke, in a convulsion of rage and pain.

"Our little boat flew dancing back from the seething vortex around him, just in season to escape being overwhelmed or crushed. He now started to run. For a short time, the line rasped, smoking, through the chocks. A few turns round the loggerhead then secured it; and with oars a-peak, and bows tilted to the sea, we went leaping onward in the wake of the tethered monster. Vain were all his struggles to break from our hold. The strands were too strong, the barbed iron too deeply fleshed, to give way. So that whether he essayed to dive or breach, or dash madly forward, the frantic creature still felt that he was held in check. At one moment, in impotent rage, he reared his immense blunt head, covered with barnacles, high above the surge, while his jaws fell together with a crash that almost made me shiver; then the upper outline of his vast form was dimly seen, gliding amidst showers of sparkling spray; while streaks of crimson on the white surf that boiled in his track, told that the shaft had been driven home.

"By this time, the whole 'school' was about us; and spouts from a hundred spiracles, with a roar that almost deafened us, were raining on every side; while in the midst of a vast surface of chafing sea, might be seen the black shapes of the rampant herd, tossing and plunging, like a legion of maddened demons. The second and third mates were in the very centre of this appalling commotion.

"At length, Dick began to lessen his impetuous speed. 'Now, my boys,' cried I, 'haul me on; wet the line, you second oars-man, as it comes in. Haul away, ship-mates!—why the devil don't you haul? Leeward side—*leeward!* I tell you! Don't you know how to approach a whale?'

"The boat brought fairly up upon his broadside as I spoke, and I gave him the lance just under the shoulder blade. At this moment, just as the boat's head was laid off; and I was straitening for a second lunge, my lance, which I had 'boned' in the first, a piercing cry from the boat-steerer drew my attention quickly aft, and I saw the waist-boat, or more properly a fragment of it, falling through the air, and underneath, the dusky forms of the struggling crew, grasping at the oars, or clinging to portions of the wreck; while a pair of flukes, descending in the midst of the confusion, fully accounted for the catastrophe. The boat had been struck and shattered by a whale!

"'Good heaven!' I exclaimed, with impatience, and in a tone which I fear showed me rather mortified at the interruption, than touched with proper feeling for the sufferers; 'good heavens—hadn't they sense enough to keep out of the red water! And I must lose this glorious prize, through their infernal stupidity!' This was the first outbreak of my selfishness.

"'But we must not see them drown, boys,' I added, upon the instant; 'cut the line!' The order had barely passed my lips, when I caught sight of the captain, who had seen the accident from the quarter deck, bearing down with oar and sail to the rescue.

"'Hold on!' I thundered, just as the knife's edge touched the line; 'for the glory of old Nantuck, hold on! The captain will pick them up, and Mocha Dick will be ours, after all!'

"This affair occurred in half the interval I have occupied in the relation. In the mean time, with the exception of a slight shudder, which once or twice shook his ponderous frame, Dick lay perfectly quiet upon the water. But suddenly, as though goaded into exertion by some fiercer pang, he started from his lethargy with apparently augmented power. Making a leap toward the boat, he darted perpendicularly downward, hurling the after oarsman, who was helmsman at the time, ten feet over the quarter, as he struck the long steering-oar in his descent. The unfortunate seaman fell, with his head forward, just upon the flukes of the whale, as he vanished, and was drawn down by suction of the closing waters, as if he had been a feather. After being carried to a great depth, as we inferred from the time he remained below the surface, he came up, panting and exhausted, and was dragged on board, amidst the hearty congratulations of his comrades.

"By this time two hundred fathoms of line had been carried spinning through the chocks, with an impetus that gave back in steam the water cast upon it. Still the gigantic creature bored his way downward, with undiminished speed. Coil after coil went over, and was swallowed up. There remained but three flakes in the tub!

"'Cut!' I shouted; 'cut quick, or he'll take us down!' But as I spoke, the hissing line flew with trebled velocity through the smoking wood, jerking the knife he was in the act of applying to the heated strands out of the hand of the boat-steerer. The boat rose on end, and her bows were buried in an instant; a hurried ejaculation, at once shriek and prayer, rose to the lips of the bravest when, unexpected mercy! the whizzing cord lost its tension, and our light bark, half filled with water, fell heavily back on her keel. A tear was in every eye, and I believe every heart bounded with gratitude, at this unlooked-for deliverance.

"Overpowered by his wounds, and exhausted by his exertions and the enormous pressure of the water above him, the immense creature was compelled to turn once more upward, for a fresh supply of air. And upward he came, indeed; shooting twenty feet of his gigantic length above the waves, by the impulse of his ascent. He was not disposed to be idle. Hardly had we succeeded in bailing out our swamping boat, when he again darted away, as it seemed to me with renewed energy. For a quarter of a mile, we parted the opposing waters as though they had offered no more resistance than air. Our game then abruptly brought to, and lay as if paralyzed, his massy frame quivering and twitching, as if under the influence of galvanism. I gave the word to haul on; and seizing a boat-spade, as we came near him, drove it twice into his small; no doubt partially disabling him by the vigor and certainty of the blows. Wheeling furiously around, he answered this salutation, by making a desperate dash at the boat's quarter. We were so near him, that to escape the shock of his onset, by any practicable manoeuvre, was out of the question.

But at the critical moment, when we expected to be crushed by the collision, his powers seemed to give way. The fatal lance had reached the seat of life. His strength failed him in mid career, and sinking quietly beneath our keel, grazing it as he wallowed along, he rose again a few rods from us, on the side opposite that where he went down.

"'Lay around, my boys, and let us set on him!' I cried, for I saw his spirit was broken at last. But the lance and spade were needless now. The work was done. The dying animal was struggling in a whirlpool of bloody foam, and the ocean far around was tinted with crimson. 'Stern all!' I shouted, as he commenced running impetuously in a circle, beating the water alternately with his head and flukes, and smiting his teeth ferociously into their sockets, with a crashing sound, in the strong spasms of dissolution. 'Stern all! or we shall be stove!'

"As I gave the command, a stream of black, clotted gore rose in a thick spout above the expiring brute, and fell in a shower around, bedewing, or rather drenching us, with a spray of blood.

"'*There's the flag!*' I exclaimed; 'there! thick as tar! Stern! every soul of ye! He's going in his flurry!' And the monster, under the convulsive influence of his final paroxysm, flung his huge tail into the air, and then, for the space of a minute, thrashed the waters on either side of him with quick and powerful blows; the sound of the concussions resembling that of the rapid discharge of artillery. He then turned slowly and heavily on his side, and lay a dead mass upon the sea through which he had so long ranged a conqueror.

"'He's fin-up at last!' I screamed, at the very top of my voice. 'Hurrah! hurrah! hurrah!' And snatching off my cap, I sent it spinning aloft, jumping at the same time from thwart to thwart, like a madman.

"We now drew alongside our floating spoil; and I seriously question if the brave commodore who first, and so nobly, broke the charm of British invincibility, by the capture of the Guerriere, felt a warmer rush of delight, as he beheld our national flag waving over the British ensign, in assurance of his victory, than I did, as I leaped upon the quarter deck of Dick's back, planted my waif-pole in the midst, and saw the little canvass flag, that tells so important and satisfactory a tale to the whaleman, fluttering above my hard-earned prize.

"The captain and second mate, each of whom had been fortunate enough to kill his fish, soon after pulled up, and congratulated me on my capture. From them I learned the particulars of the third mate's disaster. He had fastened, and his fish was sounding, when another whale suddenly rose, almost directly beneath the boat, and with a single blow of his small, absolutely cut it in twain, flinging the bows, and those who occupied that portion of the frail fabric, far into the air. Rendered insensible, or immediately killed by the shock, two of the crew sank without a struggle, while a third, unable in his confusion to disengage himself from the flakes of the tow-line, with which he had become entangled, was, together with the fragment to which the warp was attached, borne down by the harpooned whale, and was seen no more! The rest, some of them severely bruised, were saved from drowning by the timely assistance of the captain.

"To get the harness on Dick, was the work of an instant; and as the ship, taking every advantage of a light breeze which had sprung up within the last hour, had stood after us, and was now but a few rods distant, we were soon under her stern. The other fish, both of which were heavy fellows, lay floating near; and the tackle being affixed to one of them without delay, all hands were soon busily engaged in cutting in. Mocha Dick was the longest whale I ever looked upon. He measured more than seventy feet from his noddle to the tips of his flukes, and yielded one hundred barrels of clear oil, with a proportionate quantity of 'head-matter". It may emphatically be said that 'the scars of his old wounds were near his new', for not less than twenty harpoons did we draw from his back; the rusted mementos of many a desperate rencounter."

The mate was silent. His yarn was reeled off. His story was told; and with far better tact than is exhibited by many a modern orator, he had the modesty and discretion to stop with its termination. In response, a glass of "o-be-joyful" went merrily round; and this tribute having been paid to courtesy, the vanquisher of Mocha Dick was unanimously called upon for a song. Too sensible and too good-natured to wait for a second solicitation, when he had the power to oblige, he took a "long pull" and a strong, at the grog as an appropriate overture to the occasion, and then, in a deep, sonorous tone, gave us the following professional ballad, accompanied by a superannuated hand-organ, which constituted the musical portion of the cabin furniture:

I.

"Don't bother my head about catching of seals!
To me there's more glory in catching of eels;
Give me a tight ship, and under snug sail,
And I ask for no more, 'long side the sperm whale,
In the Indian Ocean,
Or Pacific Ocean,
No matter *what* ocean;
Pull ahead, yo heave O!

II.

"When our anchor's a-peak, sweethearts and wives
Yield a warm drop at parting, breathe a prayer for our lives;
With hearts full of promise, they kiss off the tear
From the eye that grows rarely dim—never with fear!
Then for the ocean, boys,
The billow's commotion, boys,
That's our devotion, boys,
Pull ahead, yo heave O!

III.

"Soon we hear the glad cry of 'Town O!—there she blows!'
Slow as night, my brave fellows, to leeward she goes:
Hard up! square the yards! then steady, lads, so!
Cries the captain, 'My maiden lance soon shall she know!'
Now we get near, boys,
In with the gear, boys,
Swing the cranes clear, boys;
Pull ahead, yo heave O!

IV.

"Our boat's in the water, each man at his oar
Bends strong to the sea, while his bark bounds before,
As the fish of all sizes, still flouncing and blowing,
With fluke and broad fin, scorn the best of hard rowing:
Hang to the oar, boys,
Another stroke more, boys;
Now line the oar, boys;
Pull ahead, yo heave O!

V.

"Then rises long Tom, who never knew fear;
Cries the captain, 'Now nail her, my bold harpooner!'
He speeds home his lance, then exclaims, 'I am fast!'
While blood, in a torrent, leaps high as the mast:
Starn! starn! hurry, hurry, boys!
She's gone in her flurry, boys,
She'll soon be in 'gurry', boys!
Pull ahead, yo heave O!

VI.

"Then give me a whaleman, wherever he be,
Who fears not a fish that can swim the salt sea;
Then give me a tight ship, and under snug sail,
And last lay me 'side of the noble sperm whale;
In the Indian ocean,
Or Pacific ocean,
Not matter *what* ocean
Pull ahead, yo heave O!"

The song "died away into an echo", and we all confessed ourselves delighted with it—save and except the gallant knight of the seal-club. He indeed allowed the lay and the music to be well enough, considering the subject; but added: "If you want to hear genuine, heart-stirring harmony, you must listen to a rookery of fur seal. For many an hour, on the rocks round Cape Horn, have I sat thus, listening to these gentry, as they clustered on the shelving cliffs above me; the surf beating at my feet, while—"

"Come, come, my old fellow!" exclaimed the captain, interrupting the loquacious sealer; "you forget the evening you are to have at Santa Maria. It is three o'clock in the morning, and more." Bidding farewell to our social and generous entertainers, we were soon safely on board our ship, when we immediately made all sail to the north.

To me, the evening had been one of singular enjoyment. Doubtless the particulars of the tale were in some degree highly colored, from the desire of the narrator to present his calling in a prominent light, and especially one that should eclipse the occupation of sealing. But making every allowance for what, after all, may be considered a natural embellishment, the facts presented may be regarded as a fair specimen of the adventures which constitute so great a portion of the romance of a whaler's life; a life which, viewing all the incidents that seem inevitably to grow out of the enterprise peculiar to it, can be said to have no parallel. Yet vast as the field is, occupied by this class of our resolute seamen, how little can we claim to know of the particulars of a whaleman's existence! That our whale ships leave port, and usually return, in the course of three years, with full cargoes, to swell the fund of national wealth, is nearly the sum of our knowledge concerning them. Could we comprehend, at a glance, the mighty surface of the Indian or Pacific seas, what a picture would open upon us of unparalleled industry and daring enterprise! What scenes of toil along the coast of Japan, up the straits of Mozambique, where the dangers of the storm, impending as they may be, are less regarded than the privations and sufferings attendant upon exclusion from all intercourse with the shore! Sail onward, and extend your view around New-Holland, to the coast of Guinea; to the eastern and western shores of Africa; to the Cape of Good Hope; and south, to the waters that lash the cliffs of Kergulan's Land, and you are ever upon the whaling-ground of the American seaman. Yet onward, to the vast expanse of the two Pacifics, with their countless summer isles, and your course is still over the common arena and highway of our whalers. The varied records of the commercial world can furnish no precedent, can present no comparison, to the intrepidity, skill, and fortitude, which seem the peculiar prerogatives of this branch of our marine. These characteristics are not the growth of forced exertion; they are incompatible with it. They are the natural result of the ardor of a free people; of a spirit of fearless independence, generated by free institutions. Under such institutions alone, can the human mind attain its fullest expansion, in the various departments of science, and the multiform pursuits of busy life.

Notes

1. 'Diego Ramirez is a small island, lying s. w. from Cape Horn.
2. A whale's name is 'Dennis,' when he spouts blood.

From "Gallapagos Islands; Prizes," David Porter, *Journal of a Cruise Made to the Pacific Ocean*

(Philadelphia: Bradford and Inskeep, 1815)

INTRODUCTION

In the first year of the War of 1812, David Porter (1780–1843) was given command of the frigate *Essex* with orders to engage in commerce raiding—an activity in which a vessel of war captures unarmed merchant vessels in an effort to disrupt enemy trade. He chose to take the war into the Pacific Ocean and to assail the British whaling fleet, which he virtually destroyed. He lost his own frigate, by 1814 starting to wear out, in a combat with two other vessels off Valparaiso in Chile. Porter wrote an account of his adventures in that war, which, like Melville's *Typee* 20 years later, was roundly condemned for its supposed immorality. Porter responded with a lengthy defense in his second edition (1822)—but he still removed many offensive passages. Both editions preserve extensive accounts of the natural history of the Galapagos Islands, which lie on the equator off the coast of South America, and the Marquesas Islands, 2,000 miles south of Hawaii.

Melville encountered Porter's book in either its first or second edition and plundered it heavily in *Typee* and later in "The Encantadas." Melville found the story of Patrick irresistible, and although hugely different morally, there are fleeting parallels in the fortunes of "Fatherless Oberlus" and the survivors of the other *Essex*, the one sunk by a whale. In 1853, Melville wrote to his publisher that he had a manuscript on "Tortoise Hunting": no doubt it also owed something to Porter's *Journal.* Bercaw #563. Modern edition: Naval Institute Press, 1986.

PRIMARY SOURCE

On our first making Chatham Island, which bore, on the morning of the 17th, N.W. by N., distant about 35 miles, I supposed it to be Hood's Island, a common stopping-place for whalers; and as this was one of the islands I was desirous of examining for them, I hauled in for it, making a signal for the Barclay to do the same; but shortly afterwards discovered Hood's Island bearing west, and bore away for it. At 7 o'clock in the evening, we were abreast the anchorage place on the N.W. part of the island, which is a good shelter from the prevalent winds; and a small island which lies off forms a secure bay, where vessels lie at anchor in 12 fathoms water, clear white sandy bottom. Here wood is to be obtained, and land tortoises in great numbers, which are highly esteemed for their excellence, and are remarkable for their size, weighing from 3 to 400 weight each. Vessels on whaling voyages among these islands generally take on board from 2 to 300 of these animals, and stow them in the hold, where, strange as it may appear, they have been known to live for a year, without food or water, and, when killed at the expiration of that time, found greatly improved in fatness and flavour. Into this bay I sent lieut. Downes with a whale-boat (I had purchased from capt. Randall) properly armed, to reconnoitre, and directed him to make a signal on discovering vessels, in order that we might send in our other boats; but at 10 o'clock he returned, after having sounded the bay without seeing any. We had entertained strong hopes of meeting enemies' vessels here, but bore the disappointment better than might be expected, considering the length of time that has elapsed since we have seen one of that description. We now hauled off toward Chatham Island, and lay to for the night, as I did not think it prudent to run for Charles' Island, the next place I intended to visit, until I could have day-light, as a reef is said to exist about two leagues to the W.N.W. of Hood's Island, and one is known to lie about nine leagues to the west of it, both said to be very dangerous; and, what is calculated to render them the more so, is the violent and irregular currents, that baffle all attempts at calculation in this part of the world. In the morning I stood to the westward, with a pleasant breeze from the east, which run us, by 2 P.M., as far as the harbour of Charles' Island. On arriving opposite to it, we could perceive no vessels; but understanding that vessels which stopped there for refreshments, such as turtle and land tortoise, and for wood, were in practice of depositing letters in a box placed for the purpose near the landing-place (which is a small beach sheltered by rocks, about the middle of the bay), I dispatched lieut. Downes to ascertain if any vessels had been lately there, and to bring off such letters as might be of use to us, if he should find any. He returned in about three hours, with several papers, taken from a box which he found nailed to a post, over which was a black sign, on which was painted *Hathaway's Post-office*. There were none of them of a late date, but they were satisfactory, inasmuch as they confirmed the information we had already received, both as respected the practice of vessels touching there, and cruizing among the other islands for whales. From those papers I obtained information, that, in June last, the following British whale-ships had put in there, on their way to the island of Albemarle, where they generally cruize for a year at a time, and some for even a longer period [*Porter's list omitted*]. There were letters also from their commanders, giving information that the American ships Perseverada [*sic*], Paddock, and the Sukey, Macy, the first with 200, the latter with 150 bbls. of sperm oil, had touched there. Considering captain Macy's letter as a rare specimen of literature, I hope I shall be pardoned for giving an exact copy of it.

> June 14th 1812.
>
> Ship Sukey John Macy 7 1/2 Months out 150 Barrels 75 days from Limo No oil Since Leaving that Port. Spanyards Very Savage Lost on the Braziel Bank John Sealin Apprentice to Capt Benjamin Worth Fell from the fore top sail Yard In A Gale of Wind. Left Diana. Capt paddock 14 day Since 250 Barrels I Leave this port this Day With 250 Turpen 8 Boat Load Wood Yesterday Went Up to Patts Landing East Side. to the Starboard hand of the Landing 1 1/2 Miles Saw 100 Turpen 20 Rods A part Road Very Bad
>
> Yours Forevir
> JOHN MACY

. . . It may be seen by captain Macy's letter, that on the east-side of the island there is another landing, which he calls Pat's landing; and this place will probably immortalize an Irishman, named Patrick Watkins, who some years since left an English ship, and took up his abode on this island, built himself a miserable hut, about a mile from the landing called after him, in a valley containing about two acres of ground capable of cultivation, and perhaps the only spot on the island which affords sufficient moisture for the purpose. Here he succeeded in raising potatoes and pumpkins in considerable quantities, which he generally exchanged for rum, or sold for cash. The appearance of this man, from the accounts I have received of him, was the most dreadful that can be imagined; ragged clothes, scarce sufficient to cover his nakedness, and covered with vermin; his red hair and beard matted, his skin much burnt, from constant exposure to the sun, and so wild and savage in his manner and appearance, that he struck every one with horror. For several years this wretched being lived by himself on this desolate spot, without any apparent desire than that of procuring rum in sufficient quantities to keep himself intoxicated, and at such times, after an absence from his hut of several days, he would be found in a state of perfect insensibility, rolling among the rocks of the mountains. He appeared to be reduced to the lowest grade to which human nature is capable, and seemed to have no desire beyond the tortoises and other animals of the island, except that of getting drunk. But this man, wretched and miserable as he may have appeared, was neither destitute of ambition nor incapable of undertaking an enterprize [*sic*] that would have appalled the heart of any other man; nor was he devoid of the talent of rousing others to second his hardihood.

He by some means became possessed of an old musket, and a few charges of powder and ball; and the possession of this weapon first set into action all his ambitious plans. He felt himself strong as the sovereign of the island, and was desirous of proving his strength on the first human being who fell in his way, which happened to be a negro, who was left in charge of a boat belonging to an American ship that had touched there or refreshments. Patrick came down to the beach where the boat lay, armed with his musket, now become his constant companion, and directed the negro, in an authoritative manner, to follow him, and on his refusal snapped his musket at him twice, which luckily missed fire. The negro, however, became intimidated, and followed him. Patrick now shouldered his musket, marched off before, and on his way up the mountains exultingly informed the negro he was henceforth to work for him, and become his slave, and that his good or bad treatment would depend on his future conduct; but arriving at a narrow defile, and perceiving Patrick off his guard, the negro seized the moment, grasped him in his arms, threw him down, tied his hands behind, shouldered him, and carried him to his boat, and when the crew had arrived he was taken on board the ship. An English smuggler was lying in the harbour at the same time, the captain of which sentenced Patrick to be severely whipped on both vessels, which was put in execution, and he was afterwards taken on shore handcuffed by the Englishmen, who compelled him to make known where he had concealed the few dollars he had been able to accumulate from the sale of his potatoes and pumpkins, which they took from him; and while they were busy in destroying his hut and garden, the wretched being made his escape, and concealed himself among the rocks in the interior of the island, until the ship had sailed, when he ventured from his sculking-place, and by means of an old file, which he drove into a tree, freed himself from the handcuffs. He now meditated a severe revenge, but concealed his intentions. Vessels continued to touch there, and Patrick, as usual, to furnish them with vegetables; but from time to time he was enabled, by administering potent draughts of his darling liquor to some of the men of their crews, and getting them so drunk that they were rendered insensible, to conceal them until the ship had sailed; when, finding themselves entirely dependent on him, they willingly enlisted under his banners, became his slaves, and he the most absolute of tyrants. By this means he had augmented the number to five, including himself, and every means was used by him to endeavour to procure arms for them, but without effect. It is supposed that his object was to have surprised some vessel, massacred her crew, and taken her off. While Patrick was meditating his plans, two ships, an American and an English vessel, touched there, and applied to Patrick for vegetables. He promised them the greatest abundance, provided they would send their boats to his landing, and their people to bring them from his garden, informing them that his rascals had become so indolent of late, that he could not get them to work. This arrangement was agreed to; two boats were sent from each vessel, and hauled on the beach. Their crews all went to Patrick's habitation, but neither he nor any of his people were to be found; and, after waiting until their patience was exhausted, they returned to the beach, where they found only the wreck of three of their boats, which were broken to pieces, and the fourth one missing. They succeeded, however, after much difficulty, in getting around to the bay opposite to their ships, where other boats were sent to their relief; and the commanders of the ships, apprehensive of some other trick, saw no security except in a flight from the island, leaving Patrick and his gang in quiet possession of the boat; but before they sailed they put a letter in a keg, giving intelligence of the affair, and moored it in the bay, where it was found by captain Randall, but not until he had sent his boat to Patrick's landing, for the purpose of procuring refreshments; and, as may be easily supposed, he felt no little inquietude until her return, when she brought him a letter from Patrick to the following purport, which was found in his hut

Sir,

I have made repeated applications to captains of vessels to sell me a boat, or to take me from this place, but in every instance met with a refusal. An opportunity presented itself to possess myself of one, and

I took advantage of it. I have been a long time endeavouring. by hard labour and suffering, to accumulate wherewith to make myself comfortable, but at different times have been robbed and maltreated, and in a late instance by captain Paddock, whose conduct in punishing me, and robbing me of about 500 dollars, in cash and other articles, neither agrees with the principles he professes nor is it such as his sleek coat would lead one to expect.[1]

On the 29th May 1809, I sail from the enchanted island in the Black Prince, bound to the Marquesas.

Do not kill the old hen; she is now sitting, and will soon have chickens.

(Signed) FATHERLESS OBERLUS

Patrick arrived alone at Guyaquil in his open boat, the rest who sailed with him having perished for want of water, or, as is generally supposed, were put to death by him, on his finding the water to grow scarce. From thence he proceeded to Payta, where he wound himself into the affection of a tawny damsel, and prevailed on her to consent to accompany him back to his enchanted island, the beauties of which he no doubt painted in glowing colours; but, from his savage appearance, he was there considered by the police as a suspicious person, and being found under the keel of a small vessel then ready to be launched, and suspected of some improper intentions, he was confined in a Payta gaol, where he now remains; and probably owing to this circumstance Charles' island, as well as the rest of the Gallapagos, may remain unpopulated for many ages to come.

. . . The possession of the vessels [*captured British whalers*], besides the great satisfaction it produced, was attended by another advantage of no less importance, as it relieved all our wants except one, to wit, the want of water. From them we obtained an abundant supply of cordage, canvas, paints, tar, and every other article necessary for the ship, of all of which she stood in great need, as our slender stock brought from America had now become worn out and useless; and besides the articles necessary for the ship, we became supplied with a stock of provisions, of a quality and quantity that removed all apprehensions of our suffering for the want of them for many months, as those vessels, when they sailed from England, were provided with provisions and stores for upwards of three years, and had not yet consumed half their stock; all were of the best quality; and, were it only for the supplying our immediate wants, the prizes were of the greatest importance to us. We found on board of them, also, wherewith to furnish our crew with several delicious meals. They had been in at James' Island, and had supplied themselves abundantly with those extraordinary animals the tortoises of the Gallapagos, which properly deserve the name of the elephant tortoise. Many of them were of a size to weigh upwards of three hundred weight; and nothing, perhaps, can be more disagreeable or clumsy than they are in their external appearance. Their motion resembles strongly that of the elephant; their steps slow, regular, and heavy; they carry their body about a foot from the ground, and their legs and feet bear no slight resemblance to the animal to which I have likened them; their neck is from 18 inches to 2 feet in length, and very slender; their head is proportioned to it, and strongly resembles that of a serpent; but, hideous and disgusting as is their appearance, no animal can possibly afford a more wholesome, luscious, and delicate food than they do; the finest green turtle is no more to be compared to them, in point of excellence, than the coarsest beef is to the finest veal; and after once tasting the Gallapagos tortoises, every other animal food fell greatly in our estimation. These animals are so fat as to require neither butter nor lard to cook them, and this fat does not possess that cloying quality, common to that of most other animals; and when tried out, it furnishes an oil superior in taste to that of the olive. The meat of this animal is the easiest of digestion, and a quantity of it, exceeding that of any other food, can be eaten, without experiencing the slightest inconvenience. But what seems the most extraordinary in this animal, is the length of time that it can exist without food; for I have been well assured, that they have been piled away among the casks in the hold of a ship, where they have been kept eighteen months, and, when killed at the expiration of that time, were found to have suffered no diminution in fatness or excellence. They carry with them a constant supply of water, in a bag at the root of the neck, which contains about two gallons; and on tasting found that in those we killed on board, it proved perfectly fresh and sweet. They are very restless when exposed to the light and heat of the sun, but will lie in the dark from one year's end to the other without moving; in the day-time, they appear remarkably quick-sighted and timid, drawing their head into their shell on the slightest motion of any object; but they are entirely destitute of hearing, as the loudest noise, even the firing of a gun, does not seem to alarm them in the slightest degree, and at night, or in the dark, they appear perfectly blind. After our tasting the flesh of those animals, we regretted that numbers of them had been thrown overboard by the crews of the vessels before their capture, to clear them for action; but a few days afterwards, at day-light in the morning, we were so fortunate as to find ourselves surrounded by about fifty of them, which were picked up and brought on board, as they had been lying in the same place where they had been thrown over, incapable of any exertion in that element, except that of stretching out their long necks.

Note

1. Captain Paddock was of the society of friends, commonly called quakers.

From Francis Allyn Olmsted, *Incidents of a Whaling Voyage*

(New York: Appleton, 1841)

INTRODUCTION

Olmsted (1819–1844) was born two weeks before Melville, but unlike his northern contemporary the North Carolina native had the privilege to attend Yale University. Melville—who remarked in *Moby-Dick* that "A whale-ship was my Yale College and my Harvard"—may have had Olmsted's book as early as the writing of *Typee* (1846). Olmsted, a convalescent gone to sea, is both like and unlike Dana, who went to sea as a common sailor after suffering vision problems as a Harvard student. Olmsted had access to Thomas Beale's *Natural History of the Sperm Whale* (which he mentions in his preface), but apparently drew his account of the *Essex* entirely from "The Whale Fishery," *North American Review* 38 (January 1834), 84–115. Bercaw #528. Modern edition: C. E. Tuttle, 1969.

PRIMARY SOURCE

The loss of the whaler Essex, of Nantucket, is one of the most remarkable in the history of the Sperm Whale Fishery. A narrative of that event by Owen Chase, mate of the ship gives a vivid description of that terrific catastrophe. "I observed" (says he) "a very large spermaceti whale, as well as I could judge, about eighty-five feet in length. He broke water about twenty rods off our weather bow, and was lying quietly with his head in a direction for the ship. He spouted two or three times and then disappeared. In less than three seconds, he came up again, about the length of the ship off, and made directly for us, at the rate of about three knots. The ship was then going with about the same velocity. His appearance and attitude gave us at first no alarm; but while I stood watching his movements, and observing him but a ship's length off, coming down for us with great celerity, I involuntarily ordered the boy at the helm to put it hard up, intending to sheer off and avoid him. The words were scarcely out of my mouth before he came down upon us at full speed, and struck the ship with his head just forward of the forechains. He gave us such an appalling and tremendous jar, as nearly threw us all on our faces. The ship brought up as suddenly and violently as if she had struck a rock, and trembled for a few minutes like a leaf. We looked at each other in perfect amazement, deprived almost of the power of speech. Many minutes elapsed before we were able to realize the dreadful accident, during which time he passed under the ship, grazing her keel as he went along, came up alongside her to leeward, and lay on the top of the water, apparently stunned with the violence of the blow, for the space of a minute. He then suddenly started off in a direction to leeward. After a few moments' reflection, and recovering in some measure from the consternation that had seized us, I of course concluded that he had stove a hole in the ship, and that it would be necessary to set the pumps a-going. Accordingly, they were rigged, but had not been in operation more than one minute, before I perceived the head of the ship to be gradually settling down in the water. I then ordered the signal to be set for the other boats—at that time in pursuit of whales—which I had scarcely despatched, before I again discovered the whale apparently in convulsions, on the top of the water, about one hundred rods to leeward. He was enveloped in the foam, that his continued and violent threshing about in the water had created around him, and I could distinctly see him smite his jaws together as if distracted with rage and fury. He remained a short time in this situation, and then started off with great velocity across the bows of the ship to windward. By this time the ship had settled down a considerable distance in the water, and I gave her up as lost. I, however, ordered the pumps to be kept constantly going, and endeavored to collect my thoughts for the occasion. I turned to the boats, two of which we then had with the ship, with an intention of clearing them away and getting all things ready to embark in them, if there should be no other resource left. While my attention was thus engaged for a moment, I was roused by the cry of the man at the hatchway, 'Here he is.—he is making for us again!' I turned round and saw the whale about one hundred rods directly ahead of us, coming down with apparently twice his ordinary speed, and to me it appeared with ten-fold fury and vengeance in his aspect. The surf flew in all directions, and his course towards us was marked by a white foam of a rod in width, which he made with a continual violent threshing of his tail. His head was about half out of water, and in that way he came upon, and again struck the ship. I was in hopes, when I descried him making for us, that by putting the ship away immediately, I should be able to cross the line of his approach before he could get up to us, and thus avoid, what I knew, if he should strike us again, would be our inevitable destruction. I called out to the helmsman 'hard up,' but she had not fallen off more than a point before we took the second shock. I should judge the speed of the ship at this time, to have been about three knots, and that of the whale about six. He struck her to windward, directly under the cat-head, and completely stove in her bows. He passed under the ship again, went off to leeward, and we saw no more of him."

This dreadful disaster occurred near the equator, at the distance of a thousand miles from land. With the scanty provisions and equipments they could save from the foundering wreck, twenty men embarked in three slender whaleboats upon the mighty ocean to buffet its surging billows, with the desperate chance of being picked up by some cruiser before reaching land, which lay at such a distance, as almost to forbid a rational hope of success. One boat was never heard of afterwards, and was probably lost; with a fate scarcely more enviable, the crews of the others, experiencing the extremest misery that human nature can endure, were picked up at sea by different ships nearly two thousand miles from the scene of the disaster. So horrible was their situation, that they were forced to draw lots to decide which of their number should be killed to appease the corroding pangs of hunger.

"There have been other instances of shipwreck, caused by the shock of these leviathans. In 1807, the ship 'Union,' of Nantucket, Captain Gardner, was totally lost between Nantucket and the Azores, by a similar concussion. But no other instance is known, in which the mischief is supposed to have been malignantly designed by the assailant, and the most experienced whalers believe that even in this case, the attack was not intentional. Mr. Chase, however, could not be persuaded to think so. He says that all he saw, produced on his mind the impression of decided and calculating mischief on the part of this maddened leviathan." (North American Review.)

From Frederick Debell Bennett, *Narrative of a Whaling Voyage round the Globe*

(London: Richard Bentley, 1840)

INTRODUCTION

The Englishman Bennett (1806–1859) began his three-year whaling voyage to the Pacific on the *Tuscan* in 1833 with the explicit purpose of studying the natural history of the sperm whale. It seems to have been a sort of custom for newly minted surgeons to gain practical experience by serving as a ship's doctor on a long voyage. His approach differs from that of Thomas Beale primarily in that Bennett's voyage is narrated in the first three-quarters of the book, and the cetology and other extensive natural history is reserved for the last quarter—rather than the other way round. Melville had Bennett's work as early as 1846 and must have recalled the second volume's extensive whaling appendices when he contemplated the composition of *Moby-Dick.* Melville's debt to Bennett is simply enormous and has not yet been studied thoroughly by Melvilleans, despite more than half a century of attention. See Vincent (1949). Bercaw #60. Modern edition: New York: Da Capo Press, 1970.

PRIMARY SOURCE

Chapter III

Natural history and habits of the Cachalot—Geographical distribution of the species—Natural indications of its places of resort.

The Sperm Whale is gregarious; and usually occurs in parties, which are termed by whalers "schools "and "pods:" the former name expressing the greater, and the latter the less number of individuals congregated together. A school may contain from twenty to fifty or more Cachalots, and is composed of females, or "cows," attended by their young, and associated with at least one adult male, or "bull," of the largest size, who acts as the guardian of the herd, and who, with a devotion highly creditable to his gallantry, generally takes a defensive position in the rear, when the school is flying from danger.

The smaller associations, or "pods," consist of young or half-grown males, which have been driven from their maternal schools, but yet retain a social disposition; and sometimes of large and adult males; although these last more commonly roam the ocean singly, or in association with the herds of females. When a solitary, or "lone" Cachalot is observed, it almost invariably proves to be an old bull; and there is some reason to believe that, in correspondence with a similar habit amongst gregarious quadrupeds, the aged male frequently retires from a social, to a secluded, state of existence.

Two or more schools occasionally coalesce, and form a very large assemblage, technically distinguished as a "body of Whales." On some tracts of ocean, peculiarly favoured as their haunts, the number of Sperm Whales, seen in one large body or in many distinct schools, is beyond all reasonable conception; and could it be accurately named, would appear incredible to persons who alone consider the vast size of this animal, or who may have traversed the main for many months without noticing more than a single whale spouting in the distance. At particular times and places, however, we have seen the ocean, for several miles around the ship, strewn with a constant succession of spouts, denoting a greater number of Cachalots than, could they all have been secured, would have afforded a full cargo of oil for three or four ships.

These large assemblies sometimes proceed at a rapid pace in one determinate direction, and are said to be "making a passage;" when, although the whales evince no appearance of alarm, their speed renders them difficult to approach, unless the boats are favourably placed to intercept them in their route.

At other times they are observed to be scattered on the surface of the sea, basking or sleeping, spouting leisurely, and exhibiting every indication of being "at home," or on their feeding-ground.

Notwithstanding its unwieldy bulk, this whale is not deficient in activity. When first pierced by the harpoon, it will tow the attached boat at the rate of more than fifteen miles an hour; but this velocity of motion is the effect of extreme excitement, and does not continue long. Under ordinary circumstances of alarm, as when conscious of being pursued by enemies, its speed averages about eight or ten miles an hour: Whale-boats, propelled by both sails and oars, and a ship, having the advantage of a strong breeze, will often succeed in overtaking the whales they pursue, or, by their near approach, compel them to seek refuge in the deep. When swimming rapidly, the Cachalot moves with an easy, regular, and majestic pace; the head being much raised above the surface of the sea, and a portion of the back being occasionally exhibited, in the action of leaping. The individuals composing a retreating party will sometimes move in lines, like a troop of horse, and exert their peculiar leaping movements, descend, rise, and often even spout, simultaneously.

A large party of Cachalots, gamboling on the surface of the ocean, is one of the most curious and imposing spectacles a whaling voyage affords: the huge size and uncouth agility of the monsters, exhibiting a strange combination of the grand and ridiculous. On such occasions, it is not unusual to observe a whale of the largest size leap from the water with the activity of a salmon, display the entire of its gigantic frame, suspended at the height of several feet in the air, and again plunge into the sea with a helpless and tremendous fall, which causes the surrounding waters to shoot up in broad and lofty columns, capped with foam; whilst others of the school leap, or "breach," in a less degree; sportively brandish their broad and fan-shaped flukes in the air; or protrude their heads perpendicularly above the waves, like columns of black rock.

As long as the Cachalot continues on the surface of the sea, it casts from its nostril a constant succession of spouts; each jet following the other after an interval of ten or fifteen seconds, with a regularity highly characteristic of this kind of whale. The respiratory jet, or spout, has the appearance of a thick and white mist, composed of numerous small drops of condensed vapour. It ascends obliquely upwards and forward; seldom rises to a greater height than six or eight feet; remains suspended in the air but a short time; and is sent forth from the spiracle by one continued effort, accompanied by a rushing sound, resembling that of a moderate surf upon a smooth beach; the superior extremity of the snout being raised high above the level of the sea at each explosive effort. It is neither abruptly terminated, nor succeeded by any perceptible sound of inspiration, or "drawback," as is the case in the spouting of many of the less valued kinds of cetacea; in fact, its sound is, on the whole, so peculiar, that the practised ear of the whaler can detect by it the close vicinity of the Cachalot, although the darkness of night should conceal the animal from his view. During a close encounter with this whale, the latter often spouts into the boats and amongst the crew, when the exhaled fluid has been observed to be foetid in odour, and to produce an acrid effect upon the skin.

When about to descend, the Cachalot assumes a vertical posture, raising its flukes *perpendicularly* in the air, an evolution which is performed leisurely, and which is regarded by whalers as a principal point for distinguishing this from most other kinds of cetacea. It is one, however, which is not invariably exhibited; for, when desirous of a sudden and temporary descent, the whale will occasionally sink in the horizontal position, or, as it is technically expressed, "settle down."

A necessity for respiring atmospheric air does not permit the Sperm Whale to continue below the surface of the water beyond a limited period: one hour is perhaps the average time for an individual of the largest size. On three occasions, when large and solitary whales were noticed, under circumstances favourable for observations in this point, I found that one Cachalot continued below the surface fifty-five minutes; a second, forty-five minutes; and the third, one hour and a half; but whatever time the whale may remain under water, after making a formal descent, it subsequently adheres to the same period with so much regularity, that its "risings" may be timed by a watch with considerable accuracy.

School Whales are less open to investigation in this respect; for, while some are rising to spout, others are descending, and it is difficult to identify individuals. A young male, harpooned by our boats, continued under water, and with the line attached to it, half an hour. The time they spend in communication with the air is very variable; when making a rapid passage, or when pursued, they swim for a long time on the surface of the water, where their progress is less impeded.

Their ordinary food is the cuttle-fish, or "squid," (*Sepia*,) many kinds of which are rejected from the stomach of the whale when the latter is attacked by the boats, as well as after death, and during the process of removing the blubber. Amongst the examples thus rejected we find solid masses of enormous size and weight, though evidently but mere fragments of the body of some vast cuttle-fish, and probably of that kind termed the "gigantic squid"—an animal with which naturalists are but little acquainted; but which has been known to attain such huge dimensions, that, when spread out beneath the surface of the sea, it has alarmed navigators by its resemblance to a reef, or shoal. We cannot fail to be impressed with a truly magnificent idea of the profusion of animal life which must necessarily exist in the ocean's depths, though invisible to man, when we thus behold creatures of such stupendous magnitude as the Sperm Whale, roaming the seas in very large assemblies, and all replete with food and of unwieldy corpulence.

It is probable that the Cachalot occasionally indulges in other food than the nutritious and helpless Sepia. While we were cruising in the North Pacific, in the year 1835, a School Whale, upon being attacked by the boats, rejected from her stomach a *bony fish*, which was secured and put into my possession. It was rather more than a foot long, of deep and compressed form, and covered with rough and silvery scales. It was uninjured, beyond some slight ravages of digestion, which had impaired the surface of the body and removed the membranes of the fins. In its stomach were small fragments of cuttle-fish. As it was not easy to conceive, that the Cachalot could take a fish of this size and structure by direct pursuit, I felt inclined to the opinion, that it had been accidentally received within the mouth of the whale, and swallowed together with the large portions of cuttle-fish with which it was also rejected. Some whalers assert, that they have seen Cachalots throw up rock-cod, and even sharks.

Owing to the great projection of the snout beyond the lower jaw, it may be requisite for this whale to turn on its side or back to seize its more bulky prey—a supposition which is strengthened by the fact, that when the animal attacks a boat with its mouth, it invariably assumes a reversed posture, carrying the lower jaw above the object it is attempting to bite.

In common with most terrestrial animals which herd together in great numbers, Sperm Whales are naturally timid, and disposed to fly from the remotest appearance of danger: and although many instances occur amongst them of a bold and mischievous disposition, which leads them, when molested, to attack and destroy both boats and men, yet such traits rather belong to the individual, than to the general character, and may be compared to the aberrations from a mild to a vicious temper, occasionally displayed amongst oxen, horses, deer, and other *herbivora.* A shoal of Dolphins, leaping in their vicinity, is sufficient to put to flight a large party of Cachalots; and when on a well-beaten cruising ground, where the whales are exceedingly watchful, the whaler finds it necessary to be constantly on his guard, not to excite or confirm their suspicions, until he has made sure of his prizes.

The signs they exhibit of a suspicion of danger, are occasionally ceasing to spout; lying motionless on the water; (evidently listening;) sweeping their flukes slowly from side to side, in search of some invisible but dreaded object; and turning upon the side, or rearing the head perpendicularly above the waves, to bring surrounding objects upon the axis of vision. When pursued, they may be considered to exhibit two degrees of alarm; namely, that which puts them to the top of their speed, and which often enables them to escape the boats; or a more powerful and overwhelming impression, produced by the near approach of their enemies, or by one of their number being injured; when they will occasionally crowd together, stationary and trembling, or make but confused and irresolute efforts to escape. The females, when attacked, will often endeavour to assist each other, and those that are uninjured will remain for a long time around their harpooned companions; while the males, in which the social disposition is less strong, as commonly make a rapid retreat, without any concern for the fate of their comrades.

It is a confirmed fact, and one often noticed with surprise by southern whalers, that upon a Cachalot being struck from a boat, others, many miles distant from the spot, will almost instantaneously express by their actions, an apparent consciousness of what has occurred, or at least of some untoward event, and either make off in alarm, or come down to the assistance of their injured companion. But, without attributing to the Cachalot an extraordinary acuteness of sight or hearing, or any more mysterious sensibility, we may perhaps find, that the violent agitation of the sea, produced by the plunges of the harpooned whale, and the more rapid and distinct conveyance of sound in water than in air,[1] are sufficient to account for the above phenomenon.

From the abundance of calves, accompanying the schools, and the great number of Sperm Whales yet visible, notwithstanding the incessant slaughter to which they have been exposed for nearly a century and a half, we are justified in believing that this species is peculiarly prolific. Like other cetaceans, they couple *more hominum*: in one instance, which came under my notice, the position of the parties was vertical; their heads being raised above the surface of the sea.

Nothing satisfactory is known about the duration of pregnancy in this whale. We observed them copulating in August, but it is probable that no particular breeding season obtains, as we observed sucking calves during the entire eight months in the year we were engaged in cruising. The female produces one at a birth, but occasionally twins, as is usual with uniparous animals. She brings forth her young in the open ocean, and indiscriminately, on whatever spot the school may chance to be at the time of her parturition. The calf accompanies the school as soon as it is born. During a chase, it was often exceedingly interesting to observe sucklings, apparently but a few days old, leaping actively and spouting high by the side of their dams, and keeping up wonderfully well with the rapid pace of the retreating party.

Intelligent whalers, who have occasionally seen the female Cachalot in the act of suckling her young, agree very closely in their descriptions of this process. They state, that the mother reposes upon her side, with the pectoral fin raised above the surface of the sea, while the calf, which is thus enabled to retain its spiracle in the air, receives the protruded nipple within the angle of the mouth—a part where it is reasonable to suppose that the tongue would also be found of some assistance.

The male Cachalot takes no part in the care of his offspring; and I have reason to believe, that the female is less remarkable for maternal affection than the Greenland whale is represented to be. When pursued by boats, the mother whale will keep in the rear of the school, retarding her pace to accommodate the calf which leaps by her side, and thus far expose herself to some danger for its sake; but other facts which I observed, are rather opposed to the opinion that she will make any strenuous or romantic efforts in defence of her young, or sacrifice her life for its safety. The whaler does not consider that a mother Cachalot is more troublesome or dangerous to destroy than one without charge; and although he will harpoon a calf, in the hope of arresting the flight of the school, or of securing the mother, the measure often fails in its object.

I believe that we have seen the female purchase her own safety by the desertion of her young. On several occasions our boats destroyed a sucking calf, in the midst of a school, without sufficient interference on the part of the mother to lead to her being identified by the harpooner. In one instance, the boats, while pursuing a school, killed a calf with a single lance-wound; the body of the little whale remained floating on the water—but none of the adults discontinued their flight. On another occasion we noticed a herd of females make off with great rapidity, and leave a small calf, swimming in an anxious and bewildered manner, in the vicinity of the ship; it continued thus deserted for some time; but was ultimately rejoined by the mother, when they both set off to overtake the main body of fugitives.

The senses of the Sperm Whale are not despised by the whaler, nor disregarded in his tactics. He considers that, in this animal, the power of vision is greater than that of hearing; and although he is ever anxious to approach his prey quietly, his principal attention is directed to avoid bringing his boat within the range of vision, or, as he expresses it, "upon the eye" of the whale. To approach in a direct line with the snout or the tail is not difficult. It is true that the splashing of oars, and the rush of a boat through the water, appear calculated to excite alarm in any animal endowed with the most ordinary power of audition, but, independent of the valvular conformation of the ear, it is probable that while the whale pursues its course through the sea, the breaking of the waves, and the noise of its own respiration, tend much to disguise other sounds.

The habitat of the Sperm Whale is more peculiarly the central and fathomless waters of oceans, or the vicinity of the most abrupt coasts. Occasionally, though very rarely, they frequent the shallower seas, so commonly the resort of the True Whale.[2] The geographical range of the species must be regarded as very extensive; since no part of the aqueous globe, excepting the Polar Seas, would appear to be altogether inimical to their habits or free from their visits. In the southern hemisphere, their excursions are known to extend as high as between the sixtieth and seventieth parallels of latitude, or off the shores of the South Shetland Islands. In the northern, their boundary can be fixed, with more accuracy, at about the sixtieth degree of latitude, for, although the species has been observed on the south coast of Greenland, and as high as the corresponding latitude off the American and Asiatic continents, it is unknown to the whaler in the Arctic Ocean.

In all the intermediate climates Sperm Whales have been found more or less abundant. They have been noticed in the Mediterranean Sea, and have been occasionally cast ashore upon many parts of the coast of continental Europe, and Great Britain.[3] Living examples have been captured in the British Channel, and one individual in the river Thames. It is, however, in the warmer seas, within, or upon the verge of, the tropics, that the Cachalot is sought with the greatest success; as in those corresponding to the intertropical coasts of Africa, America, Asia, and New Holland, or surrounding the Indian and Polynesian Islands; but more especially and uniformly in the "line currents," which extend from the equator to about the seventh degree of north and south latitude, both in the western and eastern hemisphere.

Large parties of these whales have been remarked to affect particular spots at distinct times; though it has not been ascertained that their visits to any one tract of water are sufficiently regular to be considered as seasonable; and it should be observed, that when *seasons* are spoken of in this fishery, the term applies less to the periodic presence of whales in the regions referred to, than to the convenience of whale ships, in regard to weather, &c. Thus the "Japan cruise," in the Pacific, extending from 20° to 40° north latitude, commences with the spring months, April and May, and concludes with the autumnal months, September and October, or when the accession of inclement weather may be expected in that region; although there is no reason to doubt, that whales continue in its waters during the entire year. While driven by winter weather from the north, the whaler occupies his time in the more genial climates of Southern America and the Equator, or "takes the off-shore season." The same remark applies, also, to the Sperm Fishery in the Indian seas; where its seasons are equally influenced by the variations of the monsoons. Nevertheless, the natives of the Society Islands expect to see the Cachalot around their shores only during the months of April and May, or, as they express the time, "when the *Vis* (Hog-plums) are ripe;" and whalers consider, that from September to December are the most profitable months for cruising on the Equator.

Vast tracts of ocean may be cruised over without the slightest trace of the Sperm Whale being perceived, whilst other, and often very limited, extents of water will exhibit the species in great abundance. Much of this apparent caprice, however, depends upon natural causes. Powerful currents, or the space intervening between two currents setting in opposite directions, are the favourite resorts of this whale, and, doubtless, where its food is found in greatest abundance, amidst the inhabitants of the deep swept together by these local streams. Hence, wheresoever currents are denoted by their concomitant marine animals, as the floating shellfish *Janthina*, *Hyalaea*, and *Cleodora*; or other mollusks, as Sea Lizards (*Glaucus*), *Velellae*, *Porpitae*, &c., with myriads of medusae, forming what whalers term "thick water," Cachalots may reasonably be expected; and their appearance often coincides in a remarkable manner with the presence of such natural indications.

Nor is it difficult to conceive, that since the inhabitants of the ocean, like those of the land, prey each upon the other—the smaller and less organized being the food of the larger and more perfect animals—so the base of this column is formed by the mollusks, simple in their structure, and much the sport of the element they inhabit, and the capital by the cetacea, and the largest or most rapacious fish.

Notes

1. Chladni estimates the velocity of sound in water to be four or five times greater than in air. The experiments of Dr. Franklin tend to prove, that sound, after travelling above a mile through water, loses but little of its intensity.

2. So great is the difference in habitat between the two cetaceans, that during the entire period we were cruising in quest of the Cachalot, we in no single instance saw an example of the True-Whale. (*B. mysticetus.*)

3. Captain T. Stavers informs me, that he has lowered his boats in pursuit of a school of Cachalots when but forty-eight hours' sail from Portsmouth harbour.

Chapter VI

Comparison between the disposition and weapons of the Greenland-whale and Cachalot—Mischievous temper often displayed by the latter whale—Its modes of defence and their frequently fatal effects to whalers—Melancholy instances—Fighting whales—Deplorable fate of the American South-Seaman Essex and her crew—Instances of individual Cachalots notoriously dangerous to attack—Accidents to boats, independent of a vicious temper on the part of the whales they attack—Forbearance of sharks towards wrecked whalers.

The amount of hazard incurred in the Greenland and Sperm Whale Fisheries may be considered as nearly equal: the difficulties attending upon an inhospitable climate in the one, being in a great measure counterbalanced by the superior activity and general powers of the whale pursued, in the other. The

True-Whale of the Arctic and Southern Seas is, as is well known, a gentle and inoffensive creature, which seldom, if ever, exhibits a decidedly combative temper, and whose only defensive organ is the tail; the Cachalot, on the other hand, is not only better armed than the True-Whale, in possessing a formidable weapon at either extremity of its enormous body, but also more frequently displays a disposition to employ those weapons offensively, and in a manner at once so artful, bold, and mischievous, as to lead to its being regarded as the most dangerous to attack of all the known species of the whale tribe.

Although the Cachalot, when first harpooned, strikes violently with its tail, plunges convulsively, and would appear to threaten destruction to every object in its vicinity, yet these actions are unpremeditated, and awkward, and can only be regarded as instinctive expressions of pain and alarm, which the whaler expects to observe, and which he is well prepared to meet; consequently, with ordinary precautions, the boats are seldom injured at this stage of their proceedings; although it is commonly believed to be the most critical and hazardous. But should the animal be allowed time to rally, it often becomes truly mischievous. Actuated by a feeling of revenge, by anxiety to escape its pursuers, or goaded to desperation by the weapons rankling in its body, it then acts with a deliberate design to do mischief; and but too frequently succeeds, either through the inevitable nature of its attacks, or through the temerity of the whaler.

Some of these whales, when attacked, will retreat but little from the spot on which they are harpooned; but rather lie, and fight with their jaws and tail until life is extinct. Others, without being themselves injured, will aid an attacked companion, and from the circumstance of their actions being less watched, often succeed in doing serious injury to the boats; whilst some few individuals make wilful, deliberate, and even judicious attempts to crush a boat with their jaws, and, unless avoided or killed, will repeat their efforts until they succeed in their object.

An "under-clip," or blow received from a whale's flukes near the surface of the water, may shatter and overturn a boat, or injure the crew by the force of the concussion alone; but human life is chiefly endangered when the tail of the animal is swept rapidly through the air, and either descends upon the boat, cutting it down to the water's edge, or encounters in its *trajet* some of the crew standing up, as the headsman, or harpooner, who are destroyed and carried away by the blow; and this last is the most common, as well as the most sudden and awful calamity recorded in the fishery.

It was by a melancholy accident of this kind that an experienced and enterprising whaler, the father of our commander, lost his life, when in command of the ship Perseverance, and outward bound on a voyage to the Pacific Ocean. He was engaged in destroying a Cachalot, on the Brazil Bank, when a rapid and inevitable blow from the flukes of the animal struck him out of the boat; his body floated on the water, and was immediately rescued and conveyed to the ship; but, although no external marks of injury were any where visible, all attempts to restore animation were of no avail, for life was totally extinct. One of the crew, pulling an oar in the same boat, was also killed by the same blow. The whale, after thus dealing destruction amongst its pursuers, effected an escape; but there is reason to suppose, from the clue of marked harpoons left in its body, that it was subsequently destroyed by an American whale-ship.

Captain T. Stavers, of the Tuscan, when cruising in the North Pacific, during the season of 1831, had the misfortune to lose his chief mate, Mr. Young, under circumstances very similar to the preceding. On the morning of the 30th of August, a small party, or "pod," of Sperm Whales was noticed from the ship, and the commander and second-mate, lowered their boats in pursuit, leaving Mr. Young on board, in charge of the vessel. While engaged in destroying a large whale, the boat of the second-mate was so severely shattered, that the consort boat was compelled to receive both the wrecked crew and the harpoon-line. The chief-mate, on observing this dilemma, lowered his boat and came to their assistance. The harpooned whale was then spouting blood and much exhausted; while a loose Cachalot, of equal size, remained in its vicinity, striking at the boats with his flukes, with the evident intention of assisting his wounded comrade. The boats were close together, and Captain Stavers had but just remarked to his mate, that as the whale was nearly dead he would leave him to complete its destruction whilst he harpooned the loose Cachalot, when the tail of the latter passed, with the rapidity of lightning, over, and in front of his boat, and simultaneously, Mr. Young, though a large and strong man, was seen flying through the air at a considerable height, and to the distance of nearly forty yards from the boat, ere he fell into the water, where he remained floating motionless on the surface for a few moments, then sank, and was seen no more. There can he no doubt that his death was instantaneous. A native of the Society Islands plunged into the water immediately the accident occurred, and endeavoured to save the body of his unfortunate officer, but it had sunk before he could swim to the spot where it fell. No injury was sustained by any other person in the boat; nor was the boat itself injured, beyond a portion of the bow being broken off, and the thigh-board, which was torn from its place and accompanied the body of the unfortunate mate, so powerful was the impulse it had received. As is customary in cases of serious accident, the line was cut from the whale, that the boats might he at liberty to render every assistance; but when it was found that no human aid could avail in this instance, the boats renewed their attack on the harpooned whale, which was soon after killed and taken to the ship, whilst the mischievous Cachalot made off, after he had been pierced with many lance-wounds. The chief mate of the British South-Seaman Lyra,[4] when in the cruising ground of Japan, in 1832, was also swept from his boat and destroyed by a blow from a whale; and similar casualties are too numerous, and uniform in their results, to permit a more particular notice.

Some Sperm Whales appear reluctant to employ their tail when attacked, but prove active and dangerous with their jaws. Such individuals often rather seek than avoid the attacking boats, and, rushing upon them with open mouth, employ every possible art to crush them with their teeth, and, if successful, will sometimes continue in the neighbourhood, biting the wreck and oars into small fragments. When thus threatening a boat, the whale usually turns, and swims upon its back, and will sometimes act in a very sluggish and unaccountable manner, keeping its formidable lower jaw suspended for some moments over the boat, in a threatening attitude, but ultimately rolling to one side, and closing its mouth harmlessly; nor is it rare to observe this whale, when pursued and attacked, retain its mouth in an expanded state for some minutes together. Such threatening demonstrations of the jaw, as well as some others with the flukes, occasionally compel a boat's crew to leap into the water, and support themselves by swimming or clinging to oars until the danger has passed.[5]

In the year 1835, the ship Pusie Hall encountered a fighting whale, which after injuring and driving off her four boats, pursued them to the ship, and withstood for some time the lances hurled at it, by the crew, from the bows of the vessel, before it could be induced to retire: in this affair a youth in one of the boats was destroyed by a blow from the whale, and one of the officers was severely lacerated by coming in contact with the animal's jaw.

A highly tragical instance of the power and ferocity occasionally displayed by the Sperm Whale, is recorded in the fate of the American South-Seaman Essex, Captain G. Pollard. This vessel, when cruising in the Pacific Ocean, in the year 1820, was wrecked by a whale under the following extraordinary circumstances. The boats had been lowered in pursuit of a school of whales, and the ship was attending them to windward. The master and second-mate were engaged with whales they had harpooned, in the midst of the school, and the chief mate had returned on board to equip a spare boat, in lieu of his own, which had been broken and rendered unserviceable. While the crew were thus occupied, the look-out at the mast-head reported that a large whale was coming rapidly down upon the ship, and the mate hastened his task, in the hope that he might be ready in time to attack it.

The Cachalot, which was of the largest size, consequently a male, and probably the guardian of the school, in the meanwhile approached the ship so closely, that although the helm was put up to avoid the contact, he struck her a severe blow, which broke off a portion of her keel. The enraged animal was then observed to retire to some distance, and again rush upon the ship with extreme velocity. His enormous head struck the starboard bow, beating in a corresponding portion of the planks, and the people on board had barely time to take to their boat, before the ship filled with water and fell over on her side. She did not sink, however, for some hours; and the crew in the boats continued near the wreck until they had obtained a small supply of provisions, when they shaped a course for land; but here, it is to be regretted, they made a fatal error. At the time the accident happened they were cruising on the Equator, in the longitude of about 118° West, with the Marquesan and Society Islands on their lee, and might have sailed in their boats to either of those groups in a comparatively short time. Under an erroneous impression, however, that all those lands were inhabited by an inhospitable race of people, they preferred pulling to windward for the coast of Peru, and in the attempt were exposed for a lengthened period to extreme privations.

The few of the crew who survived their complicated disasters first made the land at Elizabeth, or Henderson's Island, a small and uninhabited spot in the South Pacific, and which until then had never been visited by Europeans. After a short continuance here, part of the survivors again put to sea in search of inhabited land, and ultimately reached the coast of South America; when an English South-Seaman sailed from Valparaiso, and rescued those of the sufferers who had been left to support a precarious existence on Elizabeth Island. By a strange fatality, Captain Pollard, who was amongst the number of survivors, had the misfortune to lose the ship he next commanded, by running her upon a coral reef (then but little known) in the North Pacific. He returned to the United States, dispirited by his ill fortune, and, engaging in agricultural pursuits, ceased to tempt any further the perils of the deep.

A few Cachalots have been noted individually as animals dangerous to attack. One was thus distinguished on the cruising ground off the coast of New Zealand, and was long known to whalers by the name of "New Zealand Tom." He is said to have been of great size; conspicuously distinguished by a white hump; and famous for the havoc he had made amongst the boats and gear of ships attempting his destruction. A second example, of similar celebrity, was known to whalers in the Straits of Timor. He had so often succeeded in repelling the attacks of his foes as to be considered invincible, but was at length dispatched by a whaler, who, forewarned of his combative temper, adopted the expedient of floating a cask on the sea, to withdraw his attention from the boats; but notwithstanding this *ruse*, the animal was not destroyed without much hard fighting, nor until the bow of one of the boats had been nipped off by his jaws.

By the line becoming entangled, or impacted, when the harpooned whale is descending rapidly, the boat may be instantaneously drawn under water, and before any measures can be used to free it. The axe, provided for the purpose of cutting the line, is, by some prudent officers, held in the hand while the line is running; and under some circumstances, nothing short of such precaution can save the boat's crew from a watery grave. Several distressing instances have occurred of the sudden and total disappearance of a boat, which had last been seen attached to a whale; much probability existing, that its loss was attributable to this accident.

Some minor mischances with the line endanger individuals, though not the entire boat's crew. A few years ago, Mr. Wilson, the third-mate of the Melantho, met his death in the Eastern Archipelago, by the harpoon-line getting suddenly displaced and carrying him out of the boat. The line was instantly cut from the whale, and for some moments the unfortunate man was seen floating, free from entanglement, at a little distance beneath the surface of the sea, but, while endeavours were making to reach him, he sunk. A large shark which accompanied the boat was observed to follow him in his descent, and he was not again seen.

The ship Seringapatam, while cruising in Timor Straits, in the season of 1836, lost a man under the following circumstances. A boat, occupied in killing a whale, received a severe blow from the flukes of the animal, which passed so close to a boy pulling the after-oar as to graze his breast slightly; while the man at the tub-oar was either cast out of the boat by the same contact or carried away by the harpoon-line. His body was found entangled in the line, when the latter was hauled into the boat, but life was extinct. It was noticed at the time, that the concussion the boat received from the whale had cast some coils of line over the shoulders of the deceased, and it is probable that he was hurried from the boat when the whale again took the line; but his disappearance was so instantaneous that, in the confusion of the moment, neither his absence nor the events connected with it were immediately noticed by his companions.

A boat may be much injured, and even destroyed, by a Sperm Whale, as the result of pure accident. This occurred to the South-Seaman Arabella, when cruising off the Society Islands, in 1836. A whale "milled," or turned suddenly round, upon receiving the harpoons, and, in his efforts to escape, struck the boat with his head, broke it into two portions, and literally swam through it. The crew escaped any serious injury, but the boat was shattered beyond repair. In the same manner a whale, rising suddenly to the surface of the water, may encounter a boat with its head or hump, and do considerable mischief without any vicious design.

It is a somewhat curious fact, that notwithstanding the myriads of sharks which assemble during the pursuit and cutting in of a Sperm Whale, it but seldom if ever occurs that whalers receive any personal injury from their attacks, although their disasters so frequently plunge them into the sea, and at times when these dangerous fish are not only numerous around them, but also display a most active and ferocious disposition. It would almost appear that "a fellow feeling makes them wondrous kind," and that they possess a distinct perception of the object the whaler has in view, and join with him in seeking to make a common prey of the persecuted Cachalot. That the prospect of a more inviting food will lead these fish to disregard man, is evinced in the well known fact, that the negroes of the Bahama Islands will, when employed in cutting up a stranded whale, enter the water and work amidst a crowd of sharks, which, eager to obtain a share of the whale, will pay no attention to the men, who they would otherwise attack and devour.

Notes

4. Formerly a man-of-war, and consort of the Alceste.

5. Upon speaking the American ship Augusta, in the South Pacific, in 1836, we found on her deck a boat, which had shortly before been nipped completely asunder by the jaws of a harpooned whale. Fortunately, none of the crew were injured; and the whale, which was spouting blood at the time it did the mischief, was soon after dispatched by the other boats.

From Charles H. Barnard, *A Narrative of the Sufferings and Adventures of Capt. Charles H. Barnard, in a Voyage round the World*

(New York: J. P. Callender, 1836)

INTRODUCTION

Although Barnard (1781–1840) has been described as a whaler, he was clearly after seal on the 1812 voyage that initiates this book. As this passage makes clear, though, it is no accident that a vessel after one animal might be perfectly well equipped to take another—removing some of the doubts about occasions like those reported fictionally in James Fenimore Cooper's *The Pilot* (1824) and *The Sea Lions* (1849). Melville borrowed this or an earlier edition of Barnard's book from Evert Duyckinck in 1848. The earlier edition of Barnard's book was published in 1829; it has not been determined which one Melville read.

Through the perfidy of shipwrecked British sailors whom Barnard offered to rescue at the outbreak of the War of 1812, Barnard's own sealing vessel, the *Nanina,* was taken over and he and his men were abandoned on the Falkland Islands. Their plight may have been exacerbated by Barnard's own irascible temperament. See Betsy Tyler, "Edmund Fanning I Presume: Or How to Flesh Out a Life through Local Records," *Historic Nantucket* 53:2 (Spring 2004), web. Sealts #38; Bercaw # 46. Modern Edition: Wesleyan University Press, 1979.

PRIMARY SOURCE

We discovered a school of spermaceti whales, succeeded in killing two, getting them alongside, cutting their blubber up before night, and setting the try-works in operation. At the same time we discovered a ship to the southward; at dusk we lost sight of her, but she kept in view of the light of our try-works, and about twelve at night she was almost within hail, when she proved to be the Cyrus, of London, Captain Davie, who, from our light, knew that we were a whaler, like himself.

In the morning we discovered a large school of whales; both ships gave chase, the wind blowing fresh, with much sea, and we succeeded in getting a very large one, which Captain Buckle calculated would make seventy barrels of oil; but the wind increasing, and bringing in a heavy sea, we lost one half of it before we could cut it up. We saw several columns of smoke rising from the Island of St. Mary's; supposed them to be smugglers' signals, and put away before the wind.

From Charles Darwin, *Journal of Researches into the Natural History and Geology of the Countries Visited during the Voyage of H.M.S.* Beagle

(New York: Harper, 1846)

INTRODUCTION

Darwin (1809–1882) is the most famous of seagoing scientists. The *Beagle* was a survey ship engaged in charting South America; Darwin joined her for her second voyage, under Robert Fitzroy (1805–1865), himself an extremely young officer for such a command. Darwin was invited as Fitzroy's intellectual companion: the latter presented Darwin with the first volume of Charles Lyell's (1797–1875) *Principles of Geology* (1830; the only volume published so far) at the start of the voyage, a book from which Darwin (who thought he was headed for a life as a conventional churchman) would learn to reconsider the age of the earth. The *Beagle*'s voyage took place from 1831 to 1836—longer than most whaling voyages—but Darwin spent only a year and a half of this time actually at sea. The *Beagle* returned to England via the Indian Ocean, thus circumnavigating the globe and giving Darwin the opportunity to visit, among other places, the Galapagos Islands. Darwin kept extensive and detailed records of the natural history he encountered, and it was in large part the evidence of these observations that provided him with the most famous examples—Darwin's finches—of the theory of natural selection he published in *The Origin of Species* in 1859.

Darwin's narrative of the voyage was published in 1839; Melville purchased his copy of *Journal of Researches* in 1847. Melville maintained a lifelong interest in the work of Darwin and Lyell, returning to them significantly as he contrasted religion and science in *Clarel* (1876). See Karen Lentz Madison and R. D. Madison, "Darwin's Year and Melville's 'New Ancient of Days,'" *America's Darwin: Darwinian Theory and U.S. Literary Culture*, ed. Tina Gianquitto and Lydia Fisher (Athens: University of Georgia Press, 2014), 86–103. Sealts # 175; Bercaw #191. Modern edition: Hafner, 1952.

PRIMARY SOURCE

In the sea around Tierra del Fuego, and at no great distance from the land, I have seen narrow lines of water of a bright red colour, from the number of crustacea, which somewhat resemble in form large prawns. The sealers call them whale-food. Whether whales feed on them I do not know; but terns, cormorants, and immense herds of great unwieldy seals derive, on some parts of the coast, their chief sustenance from these swimming crabs. Seamen invariably attribute the discoloration of the water to spawn; but I found this to be the case only on one occasion. At the distance of several leagues from the Archipelago of the Galapagos, the ship sailed through three strips of a dark yellowish, or mud-like water; these strips were some miles long, but only a few yards wide, and they were separated from the surrounding water by a sinuous yet distinct margin. The colour was caused by little gelatinous balls, about the fifth of an inch in diameter, in which numerous minute spherical ovules were embedded: they were of two distinct kinds, one being of a reddish colour and of a different shape from the other. I cannot form a conjecture as to what two kinds of animals these belonged. Captain Colnett remarks that this appearance is very common among the Galapagos Islands, and that the direction of the bands indicates that of the currents; in the described case, however, the line was caused by the wind. The only other appearance which I have to notice, is a thin oily coat on the water which displays iridescent colours. I saw a considerable tract of the ocean thus covered on the coast of Brazil; the seamen attributed it to the putrefying carcass of some whale, which probably was floating at no great distance. I do not here mention the minute gelatinous particles, hereafter to be referred to, which are frequently dispersed throughout the water, for they are not sufficiently abundant to create any change of colour.

There are two circumstances in the above accounts which appear remarkable: first, how do the various bodies which form the bands with defined edges keep together? In the case of the prawn-like crabs, their movements were as coinstantaneous as in a regiment of soldiers; but this cannot happen from anything like voluntary action with the ovules, or the confervae, nor is it probable among the infusoria. Secondly, what causes the length and narrowness of the bands? The appearance so much resembles that which may be seen in every torrent, where the stream uncoils into long streaks the froth collected in the eddies, that I must attribute the effect to a similar action either of the currents of the air or sea. Under this supposition we must believe that the various organised bodies are produced in certain favourable places, and are thence removed by the set of either wind or water. I confess, however, there is a very great difficulty in imagining any one spot to be the birthplace of the millions of millions of animalcula and confervae: for whence come the germs at such points?—the parent bodies having been distributed by the winds and waves over the immense ocean. But on no other hypothesis can I understand their linear grouping. I may add that Scoresby remarks that green water abounding with pelagic animals is invariably found in a certain part of the Arctic Sea. . . .

They often suffer from famine: I heard Mr. Low, a sealing-master intimately acquainted with the natives of this country, give a curious account of the state of a party of one hundred and fifty natives on the west coast, who were very thin and in great distress. A succession of gales prevented the women from getting shell-fish on the rocks, and they could not go out in their canoes to catch seal. A small party of these men one morning set out, and the other Indians explained to him that they were going a four days' journey for food: on their return, Low went to meet them, and he found them excessively tired, each man carrying a great square piece of putrid whales-blubber with a hole in the middle, through which they put their heads, like the Gauchos do through their ponchos or cloaks. As soon as the blubber was brought into a wigwam, an old man cut off thin slices, and muttering over them, broiled them for a minute, and distributed them to the famished party, who during this time preserved a profound silence. Mr. Low believes that whenever a whale is cast on shore, the natives bury large pieces of it in the sand, as a resource in time of famine; and a native boy, whom he had on board, once found a stock thus buried. The different tribes when at war are cannibals. From the concurrent, but quite independent evidence of the boy taken by Mr. Low, and of Jemmy Button, it is certainly true, that when pressed in winter by hunger they kill and devour their old women before they kill their dogs: the boy, being asked by Mr. Low why they did this, answered, "Doggies catch otters, old women no." This boy described the manner in which they are killed by being held over smoke and thus choked; he imitated their screams as a joke, and described the parts of their bodies which are considered best to eat. Horrid as such a death by the hands of their friends and relatives must be, the fears of the old women, when hunger begins to press, are more painful to think of; we were told that they then often run away into the mountains, but that they are pursued by the men and brought back to the slaughter-house at their own firesides! . . .

In deep water, far from the land, the number of living creatures is extremely small: south of the latitude 35°, I never succeeded in catching anything besides some beroe, and a few species of minute entomostracous crustacea. In shoaler water, at the distance of a few miles from the coast, very many kinds of crustacea and some other animals are numerous, but only during the night. Between latitudes 56° and 57° south of Cape Horn, the net was put astern several times; it never, however, brought up anything besides a few of two extremely minute species of Entomostraca. Yet whales and seals, petrels and albatross, are exceedingly abundant throughout this part of the ocean. It has always been a mystery to me on what the albatross, which lives far from the shore, can subsist; I presume that, like the condor, it is able to fast long; and that one good feast on the carcass of a putrid whale lasts for a long time. The central and intertropical parts of the Atlantic swarm with Pteropoda, Crustacea, and Radiata, and with their devourers the flying-fish, and again with their devourers the bonitos and albicores; I presume that the numerous lower pelagic animals feed on the Infusoria, which are now known, from the researches of Ehrenberg, to abound in the open ocean: but on what, in the clear blue water, do these Infusoria subsist? . . .

On our return in the morning (28th) we were delighted to find all quiet, and the men employed in their canoes spearing fish. Captain Fitz Roy determined to send the yawl and one whale-boat back to the ship; and to proceed with the two other boats, one under his own command (in which he most kindly allowed me to accompany him), and one under Mr. Hammond, to survey the western parts of the Beagle Channel, and afterwards to return and visit the settlement. The day to our astonishment was overpoweringly hot, so that our skins were scorched; with this beautiful weather, the view in the middle of the Beagle Channel was very remarkable. Looking towards either hand, no object intercepted the vanishing points of this long canal between the mountains. The circumstance of its being an arm of the sea was rendered very evident by several huge whales[1] spouting in different directions. On one occasion I saw two of these monsters, probably male and female, slowly swimming one after the other, within less than a stone's throw of the shore, over which the beech-tree extended its branches.

We sailed on till it was dark, and then pitched our tents in a quiet creek. The greatest luxury was to find for our beds a beach of pebbles, for they were dry and yielded to the body. Peaty soil is damp; rock is uneven and hard; sand gets into one's meat, when cooked and eaten boat-fashion; but when lying in our blanket-bags, on a good bed of smooth pebbles, we passed most comfortable nights.

Note

1. One day, off the East coast of Tierra del Fuego, we saw a grand sight in several spermaceti whales jumping upright quite out of the water, with the exception of their tail-fins. As they fell down sideways, they splashed the water high up, and the sound reverberated like a distant broadside.

From J. Ross Browne, *Etchings of a Whaling Cruise*

(New York: Harper & Brothers, 1846)

INTRODUCTION

J. Ross Browne (1821–1875) arrived in the United States from Ireland with his family in 1833. Far from being a child of privilege like Francis A. Olmsted or Richard Henry Dana Jr., Browne went whaling as a "green hand" after failing to become a doctor and working on a riverboat. After the success of his whaling book, he became a significant journalist of the American West.

Browne supplemented his own narrative of his whaling voyage with a 70-page fine-print appendix of whaling matter gleaned from authorities later consulted by Melville—indeed, it's a toss-up whether Browne or the *Penny Cyclopædia* was more useful in provided Melville with a shortcut to cetology. Browne quoted large blocks of text, hardly synthesizing at all. He quoted or cited Beale, Hunter, Henry's *History of Great Britain*, M'Culloch's *Commercial Dictionary*, Obed Macy's *The History of Nantucket* (with complete bibliographical data), Shaw's *Natural History*, Anderson's *Historical and Chronological Deduction of the Origin of Commerce,* Grinnell, Rockwell, and Davis's addresses to Congress, Honolulu newspapers, and Wilkes's *Narrative* (including the passage on medusa included below directly from Wilkes). See Vincent (1849). Sealts #88; Bercaw #82. Modern edition: Cambridge: Harvard University Press, 1968.

PRIMARY SOURCE

October 13*th.*—"There she blows!" was sung out from the mast-head.

"Where away?" demanded the captain.

"Three points off the lee bow, sir."

"Raise up your wheel. Steady!"

"Steady, sir."

"Mast-head ahoy! Do you see that whale now?"

"Ay, ay, sir! A school of sperm whales! There she blows! There she breaches!"

"Sing out! Sing out every time!"

"Ay, ay, sir! There she blows! There—there—*thar'* she blows—bowes—bo-o-o-s!"

"How far off!"

"Two miles and a half!"

"Thunder and lightning! so near! Call all hands! Clew up the fore-t'gallant-sail—there! belay! Hard down" your wheel! Haul aback the main yard! Get your tubs in your boats. Bear a hand! Clear your falls! stand by all to lower! All ready?"

"All ready, sir!"

"Lower away!"

Down went the boats with a splash. Each boat's crew sprang over the rail, and in an instant the larboard, starboard, and waist boats were manned. There was great rivalry in getting the start. The waist-boat got off in pretty good time; and away went all three, dashing the water high over their bows. Nothing could be more exciting than the chase. The larboard boat, commanded by the mate, and the waist-boat, by the second mate, were head and head.

"Give way, my lads, give way!" shouted P—, our headsman; "we gain on them; give way! A long, steady stroke! That's the way to tell it!"

"Ay, ay!" cried Tabor, our boat-steerer. "What d'ye say, boys? Shall we lick 'em?"

"Pull! pull like vengeance!" echoed the crew; and we danced over the waves, scarcely seeming to touch them

The chase was now truly soul-stirring. Some-times the larboard, then the starboard, then the waist-boat took the lead. It was a severe trial of skill and muscle. After we had run two miles at this rate, the whales turned flukes, going dead to wind-ward.

"Now for it, my lads!" cried P—. "We'll have them the next rising. Now pile it on! a long, steady pull! That's it! that's the way! Those whales belong to us. Don't give out! Half an hour more, and they're our whales!"

The other boats had veered off at either side of us, and continued the chase with renewed ardor. In about half an hour we lay on our oars to look round for the whales.

"There she blows! right ahead!" shouted Tabor, fairly dancing with delight.

"There she blows! There she blows!"

"Oh, Lord, boys, spring!" cried P—.

"Spring it is! What d'ye say, now, chummies? Shall we take those whales?"

To this general appeal every man replied by putting his weight on his oar, and exerting his utmost strength. The boat flew through the water with incredible swiftness, scarcely rising to the waves. A large bull whale lay about a quarter of a mile ahead of us, lazily rolling in the trough of the sea. The larboard and starboard boats were far to leeward of us, tugging hard to get a chance at the other whales, which were now blowing in every direction.

"Give way! give way, my hearties!" cried P—, putting his weight against the aft oar. "Do you love gin? A bottle of gin to the best man! Oh, pile it on while you have breath! pile it on!"

"On with the beef, chummies! Smash every oar! double 'em up, or break 'em!"

"Every devil's imp of you, pull! No talking; lay back to it; now or never!"

On dashed the boat, cleaving its way through the rough sea as if the briny element were blue smoke. The whale, however, turned flukes before we could reach him. When he appeared again above the surface of the water, it was evident that he had milled while down, by which manoeuver he gained on us nearly a mile. The chase was now almost hopeless, as he was making to windward rapidly. A heavy, black cloud was on the horizon, portending an approaching squall, and the barque was fast fading from sight. Still we were not to be baffled by discouraging circumstances of this kind, and we braced our sinews for a grand and final effort.

"Never give up, my lads!" said the headsman, in a cheering voice. "Mark my words, we'll have that whale yet. Only *think* he's ours, and there's no mistake about it, he *will* be ours. Now for a hard, steady pull! Give way!"

"Give way, sir! Give way, all!"

"There she blows! Oh, pull, my lively lads! Only a mile off! There she blows!"

The wind had by this time increased almost to a gale, and the heavy black clouds were scattering over us far and wide. Part of the squall had passed off to leeward, and entirely concealed the barque. Our situation was rather unpleasant: in a rough sea, the other boats out of sight, and each moment the wind increasing.

We continued to strain every muscle till we were hard upon the whale. Tabor sprang to the bow, and stood by with the harpoon.

"Softly, softly, my lads," said the headsman.

"Ay, ay, sir!"

"Hush-h-h! softly. Now's your time, Tabor!"

Tabor let fly the harpoon, and buried the iron.

"Give him another!"

"Ay, ay! Stern all!"

"Stern all!" thundered P—.

"Stern all!"

And, as we rapidly backed from the whale, he flung his tremendous flukes high in the air, covering us with a cloud of spray. He then sounded, making the line whiz as it passed through the chocks. When he rose to the surface again, we hauled up, and the second mate stood ready in the bow to dispatch him with lances.

"*Spouting blood!*" said Tabor. "*He's a dead whale!* He won't need much lancing." It was true enough; for, before the officer could get within dart of him, he commenced his dying struggles. The sea was crimsoned with his blood. By the time we had reached him, he was belly up. We lay upon our oars a moment to witness his last throes, and, when he had turned his head toward the sun, a loud, simultaneous cheer burst from every lip

A low, rumbling sound, like the roar of a distant waterfall, now reached our ears. Each moment it grew louder. The whole expansive arch of the heavens became dark with clouds tossing, flying, swelling, and whirling over and over, like the surges of an angry sea. A white cloud, gleaming against the black mass behind it, came sweeping toward us, stretching forth its long, white arms, as if to grasp us in its fatal embrace. Louder and still louder it growled; yet the air was still and heavy around us. Now the white cloud spread, whirled over, and lost its hoary head; now it wore the mane and fore feet of a lion; now the heads of a dragon, with their tremendous jaws extended. Writhing, hissing, roaring, it swept toward us. The demon of wrath could not have assumed a more frightful form. The whole face of the ocean was hidden in utter darkness, save within a circle of a few hundred yards. Our little boat floated on a sea almost unruffled by a breath of wind. The heavy swell rolled lazily past us; yet a death-like calmness reigned in the air. Beyond the circle all was strife; within, all peace. We gazed anxiously in each other's faces; but not a word was spoken. Even the veteran harpooneer looked upon the clouds with a face of unusual solemnity, as we lay upon our oars, awed to silence by the sublimity of the scene. The ominous stillness of every thing within the circle became painful. For many long minutes the surface of the water remained nearly smooth. We dreaded, but longed for a change. This state of suspense was growing intolerable. I could bear the deep, long-drawn respirations of those around me; I saw the quick, anxious glances they turned to windward; and I almost fancied I could read every thought that passed within their breasts. Suddenly a white streak of foam appeared within a hundred yards. Scarcely had we unshipped our oars, when the squall burst upon us with a stunning violence. The weather side of the boat was raised high out of the water, and the rushing foam dashed over the gunwale in torrents. We soon trimmed her, however, and, by hard bailing, got her clear of water. It is utterly impossible to conceive the violence of the wind. Small as the surface exposed to the squall was, we flew through the foaming seas, dragging the dead body of the whale after us with incredible velocity. Thus situated, entirely at the mercy of the wind, and sea, we continued every moment to increase our distance from the barque. When the squall abated, we came to under the lee of the whale, and looked to leeward for the barque. Not a speck could be seen on the horizon! Night was rapidly approaching, and we were alone upon the broad, angry ocean!

"Ship your oars," said the headsman; "we'll not part company with old Blubber yet. If we can't make the barque, we can make land somewhere."

"Ay, ay," said Tabor, with a sly leer, "and live on roast-beef and turkey while we're making it."

With heavy hearts and many misgivings we shipped our oars, heartily wishing the whale in the devil's try-pots; for we thought it rather hard that our lives should be risked for a few barrels of oil. For two hours we pulled a long, lazy, dogged stroke, without a sign of relief. At last Tabor stood up on the bow to look out, and we lay on our oars.

"Well, Tabor, what d'ye see?" was the general inquiry.

"Why," said Tabor, coolly rolling the quid from his weather to his lee cheek, "I see a cussed old barque that looks like Granny Howland's wash-tub, with a few broom-sticks rigged up in the middle of it."

"Pull, you devils!" cried P—; "there's duff in the cook's coppers."

"Yes! I think I smell it," said Tabor.

It was nearly dark when we arrived alongside of the barque with our prize; but what was our surprise to find that the starboard and larboard boats had killed *five* whales between them!. They were all of a small size, and did not average more than fifteen barrels each.

That night not a breath of air ruffled the clear, broad ocean as it swelled beneath and around us, forming a multitude of mirrors that reflected all the beauties of the splendid canopy above. The moon arose with unusual brilliancy. It was a night for the winged spirits of the air. I enjoyed a melancholy pleasure in walking the decks beneath the soft moon-beams, thinking of past times. Silence reigned over the deep. The calm, broad ocean presented a beautiful simile of repose, and the light, shadowy clouds floated motionless in the air, as if in awe of the mighty wilderness of waters beneath them. A clear, silvery light beamed over the glassy swell; and far away the moon's rays, casting their soft and delicate glow over the whole scene, gradually vanished in a dreamy haze upon the horizon. I gazed with pensive feelings upon this scene; so calm, so heavenly, so unrivalled in its loveliness; and I thought, with a sigh, of the coming day: the fiery, tropical sun; the loud, harsh voices of the officers giving orders; the heat and smoke of the try-works; and all the realities of a whaleman's life. I have heard of the solitude of the desert; but what can compare with that of the ocean at such a time as this?

Never had the sea looked more beautiful than it did that night. It was a source of pleasure to feel that, notwithstanding the wretched life I led, there were still left a few of the better feelings of my nature. A passage in the "Vision of Don Roderic" occurred to me as singularly expressive of the checkered fortunes of a sea-farer. Well might I hope the light cloud which occasionally obscured the moon's brightness might prove a happy omen of my future fate:

> "Melting, as a wreath of snow it hangs
> In folds of wavy silver round, and clothes
> The orb in richer beauties than her own;
> Then, passing, leaves her in her light serene."

At daylight next morning all hands were called, and set to work getting up the cutting tackle, and making other preparations for cutting in. As this process of "cutting in" seems to be but imperfectly understood by those who have not been engaged in the business, it would perhaps be well enough to devote a page or two of description to it in this place.

When the whale has been towed alongside by the boats, it is firmly secured by a large rope attached to the "small" by a running noose. There is not a little ingenuity in the manner in which the fluke rope is first passed under the body of the-whale. A small line, to which a lead is fastened, with a block of wood at the extremity, several fathoms from the lead, is thrown over between the whale and the ship's side. From the impetus given to the lead, it sinks in a diagonal direction, drawing the block down after it. One end of the lead line is fastened to the end of the fluke rope on board, and the block attached to the other rises at the off side of the whale. It is then hauled on board by means of a wire hook fastened to a long pole, and, in hauling it in, the fluke rope passes round under the body of the whale, till the end arrives on board, when it is passed through the loop in the other extremity, and thus a running noose is formed, which is easily slipped down to the small. The fluke rope is then made fast on the forecastle, and the flukes are hauled up to the bow, or as near as they will reach, leaving the head pointed aft. Of course, the size of the vessel and the length of the whale make a great difference; but in general the head reaches to the quarter. To prevent concussion, the whale is always on the weather side. The progress of the vessel, which is usually under easy sail during the time of cutting in, keeps the whale from drifting out at right angles from the side; though, in most cases, the head is kept in its appropriate position by a small rope made fast aft.

The cutting tackle is attached to a powerful strap, or pendant, passing round the mast in the main-top by two large blocks. There are, in fact, two tackles, the falls of which pass round the windlass. To each of these tackles is attached a large blubber hook, which, upon being made fast to the blubber, are hauled up by the windlass, one only being in operation at a time, so that when the first strip of blubber, or "blanket piece," reaches the stationary block on the pendant, the other can be made fast by a strap and bolt of wood to a hole cut below the point at which that blanket piece is to be cut off. I have endeavored to give some idea of this part of the process in the frontispiece accompanying the work. The blanket pieces are stripped off in a spiral direction, running down toward the flukes; the whale turning, at every heave of the windlass, till the whole covering of blubber is stripped off to the flukes, which are hoisted on board, and those parts containing oil cut away, and the remainder thrown overboard. The head having, in the first place, been cut off and secured to the stern, is now hauled up, with the nose down, if too large to be taken on board, and hoisted as far out of the water as may be found convenient, and the oil or liquid spermaceti bailed out with a vessel attached to a long pole, and thus taken in and saved. As there is no little risk attending this mode of getting the spermaceti, and a great deal of waste, the head is always taken on board, when not too large or heavy.

The "case," which is the name given by whalers to the head, sometimes contains from ten to fifteen barrels of oil and spermaceti. A single "blanket piece" not unfrequently weighs a ton or upward. In hauling it up by the tackles, it careens the vessel over frequently to an angle of fifteen or twenty degrees, owing to its own great weight, combined with that of the whale, the upper surface of which it raises several feet out of the water. When the blanket piece has reached the stationary block in the top, it is cut off by a boat-steerer, who stands by with a boarding knife, having first, however, been secured below by the other blubber hook, which is hauled taught, to prevent it from breaking away by too sudden a jerk. The upper piece then swings in, and, when it ceases its pendulating motion, is dropped down into the hold or blubber-room, where it is cut up into blocks of a foot and a half or two feet in length, and eight or ten inches in width. These blocks are called "horse pieces." The white, hard blocks, containing but little oil, and which are found near the small, and at the flukes, are called "white horse." The carcass of the whale, when stripped of its blubber, is cast loose, and soon sinks from the want of its buoyant covering. I have seen it float astern, however, some distance without sinking.

Breakfast over, all hands were called to cut in. Six or eight men were stationed at the windlass, two in the blubber-room, and the boat-steerers in the waist. The first and second mates took their station on a couple of stages, or platforms, rigged out at the gangway, each provided with a spade. One of the boat-steerers, whose turn it was to fasten the blubber hook, went down over the side on the whale's back, and, after several unsuccessful attempts, and rather an uncomfortable ducking, performed his task. While yet on the whale's back, a large, hungry-looking shark, which had been eyeing him for some time very anxiously, was washed up behind him by a heavy sea, and apparently loath to lose so good an opportunity of making a meal, began to work his way along the slimy surface till within a foot or two of the boat-steerer's heels. The officers happened at the moment to be looking up at the pendant block, and in all probability the man would have been seriously injured, if not carried off bodily, but for the timely alarm of one of the crew. The mate immediately turned to see what was the matter, and perceiving the critical position of the boat-steerer, brought his spade to bear upon the shark, and at a single dart chopped off his tail.

Strange to say, the greedy monster did not appear to be particularly concerned at this indignity, but, sliding back into his native element, very leisurely swam off, to the great apparent amusement of his comrades, who pursued him with every variety of gyrations. It surprised me to see with what cool indifference the boat-steerer witnessed the whole transaction. I do not remember that he said a word about it.

The various duties being apportioned to the men without favor or choice, it fell to my lot to sit on the weather side of the quarter-deck and turn the grindstone; a tiresome and monotonous task. The cooper attended to the sharpening of the spades, boarding knives, and other implements used in "cutting in." I am not sure that I had the hardest of the work to do, but it certainly was the most unpleasant; for I could not prevail upon any of the hands to change places with me, even for a brief period. My appearance at this time would have been somewhat striking to some of my friends in Washington. With my duck frock all black with whale-gurry, my trowsers [*sic*] torn and smeared with rough work, my red Scotch cap half-way over my eyes, and my face oily and sunburned, I certainly looked as little like my original self as one can well imagine. There I turned that grindstone, and turned on hour after hour, and turned the palm of my- right hand into a great blister, and turned the palm of my left into another; turned both my arms into a personified pain; turned every remnant of romance out of my head; turned and turned till my grand tour seemed to have turned into a grindstone; round and round I turned that stone till I began to think I was a piece of the handle, and turned with it; and my head appeared to turn, and my feet to turn, and the game-legged cooper to turn, and the ship to turn, and the sea, and the whale, and the sharks, and the clouds, and all creation seemed to be turning with myself and that grindstone! Having at last contrived to get a sufficient number of spades sharpened ahead of the mates, I peeped over the quarter-rail to see how they were getting on. The sharks had by this time gathered around the vessel in immense numbers, and eight or ten were fighting just under the quarter for a piece of the whale's carcass which had been cut away. Watching my opportunity, I snatched up a spade, slipped it over while the captain was forward, and began a terrible onslaught among the sharks. With five or six thrusts I killed four of the greedy monsters, by striking them on the back of the head, and cutting the principal artery. This was quite a refreshing little episode in my business of turning; and my success in the destruction of sharks induced me to believe that I had a greater natural turn for sport than the monotonous *turning* of a grindstone. But my amusement was of short duration. The eagle eye of the captain espied me before I could get in the long pole of the spade.

"Ha! what are you at there?" cried his highness directly behind me, at the very moment when I supposed he was on the forecastle giving orders to the men. "What are you at, hey?"

"Keeping off the sharks, sir."

"Who told you to keep 'em off?"

"Nobody, sir."

"Haul in that spade directly!"

"Ay, ay, sir!"

"And, look'ee, if I catch you keepin' off any more sharks, I'll wipe you down with a rope's end!"

"Ay, ay, sir!"

"Cooper, hain't you got no work for this fellow?"

"Not just now, sir."

"Go to the windlass, then, and rest yourself on a handspike!"

"Ay, ay, sir!"

I had reason to consider my shark-killing a poor speculation. The heaving and surging at the windlass was but a questionable improvement upon my old business of turning the grindstone. At the word, "Heave away!" somebody struck up an extemporaneous song, which, to the best of my recollection, had no particular claims to poetical merit, but ran somewhat thus:

"Heave him up! O he yo!
Butter and cheese for breakfast
Raise the dead! O he yo!
The steward he's a makin' swankey.
Heave away! O he yo!
Duff for dinner! Duff for dinner!
Now I see it! O he yo!
Hurrah for the Cape Cod gals!
Now I don't. O he yo!
Round the corner, Sally!
Up she comes! O he yo!
Slap-jacks for supper!
Re-re-ra–ra-oo-we ye yo ho! Them's 'um!"

At the conclusion of this medley, the captain, who had seated himself in the starboard quarter boat to inspect the cutting, began to criticise the mate's style of cutting rather severely. Now the mate, be it known, was really a very skillful whaleman, and handled the spade with an unerring hand. The "old man's" comments, thus lowering him in the eyes of the crew, by no means pleased him.

"I say, Mr. D—," persisted the captain, "that's not the way to cut in a whale. I don't want no such work as that about me."

"It's *my* way, sir," replied the mate, getting very red in the face.

"Well, I never see a whale cut that way. I ain't used to it; I won't have it."

"You haven't seen every thing yet, sir. I've always cut whales this way, and always mean to do it."

"No you won't; not here you won't. You can just cut as I tell you."

"I reckon I know my own business; Captain A—. Now, sir, I'm not a going to be dictated to in this manner. If you think you can cut a whale better than I can, you'd better take my place."

After some more angry words, during which both the captain and mate became much excited, and threatened to whip each other, the quarrel ceased, and the "old man" went below in high dudgeon.

This was all "nuts" for me. I was rejoiced to see somebody among the privileged few talk up to him as he deserved. Indeed, I was itching to express my own personal opinion on the subject, but had no particular fancy for the mode of "wiping down" hinted at a short time previously.

Cutting in, trying out, and clearing up the decks, occupied us for the next six days. We had an average of five hours' sleep out of the twenty-four.

Working incessantly in oil, which penetrated to the skin, and kept us in a most uncomfortable condition, besides being continually saturated with salt water, produced a very disagreeable effect upon those who were not accustomed to such things, by chafing the skin, and causing painful tumors to break out over the whole body. Before I had half finished my share of the labor, I heartily wished myself in the meanest dog-kennel ashore, or, to borrow an old idea, I should have considered myself in an enviable situation had some enemy been kicking me down Pennsylvania Avenue. Tabor, the oldest whaleman on board, who laughed at hardships, and took all the disagreeable parts of his duty with the utmost good- humor, frequently joked me on my "grand tour to Europe."

"What do you think of whaling now, B—? Is it equal to traveling in Italy?"

"I think not, Tabor."

Appendix

It is hardly necessary for me to state that I make no pretensions to scientific attainments. At the time of writing the preceding narrative, I had never read a single work of any description relating to the whale fishery. Since my return from the Indian Ocean, the study of the natural history of the whale has engrossed no small share of my attention; but, wishing my journal to stand upon its own footing, I have not embraced in it the information which I might have obtained from the works of more experienced writers. As an offset, however, to a light and sketchy narrative of mere personal adventures, professing to be nothing more than a delineation of the life of a whaleman, I am induced to believe a compilation of more solid reading matter will not be altogether unacceptable. Having had access to the various works in the Library of Congress on the whale fishery, I flatter myself I shall be able to present to the reader some interesting and curious facts, not easily found elsewhere, and which may serve to satisfy his mind on many particulars not within the scope of my observation. I wish it to be borne in mind that, in making this compilation, I am indebted to the writings of others, and lay no claims to originality, being quite willing to rest my chances of approbation or censure on the foregoing narrative alone. Convinced that I can not improve what others, better versed in the subject, have bestowed so much labor upon, I shall, for the most part, give the extracts without addition or alteration.

I have found it extremely difficult to obtain any definite and authentic facts in relation to the origin of the whale fishery. The works which I have before me, containing the earliest records on this subject, are of so conflicting a character, that, after wading through volumes embracing a great variety of other topics, I am puzzled upon which to place the most reliance. Instead, therefore, of attempting any consecutive arrangement of these records, I shall lay before the reader such extracts from works of an authentic character as will enable him to judge for himself; and this I shall endeavor to do with as much regard to order as the nature of the subject will permit. It is true, I present nothing new; but I find an apology for that in the fact that it took me many weeks to collate what I do present; and no compilation of the kind having yet been made, the inquirer will, at least, be saved the trouble of ransacking more than twenty voluminous works to find what is here embraced within the limits of a few dozen pages. I am confident the rare and interesting character of the extracts will amply repay a perusal, and my only regret is, that I am unable to give them in a more methodical form.

It is admitted, I believe, that the Norwegians were the first to engage in the capture of the whale, and that they carried on a fishery long before any other European nation. As early as 887, according to Anderson (in his Historical and Chronological Deduction of the Origin of Commerce), or, as Hakluyt thinks, about 890, "our excellent King Alfred received from one Octher, a Norwegian, an account of his discoveries northward on the Coast of Norway; a coast which seems to have been before very little, if at all, known to the Anglo-Saxons. There is one very remarkable thing in this account; for he tells King Alfred 'that he sailed along the Norway coast, so far north as commonly the whale-hunters used to travel;' which shows the great antiquity of whale fishing, though undoubtedly then, and long after, the use of what is usually called whalebone was not known; so that they fished for whales merely on account of their fat or oil." Macpherson, in his improved edition of this work, in which many of the statements are corrected and valuable additions made, does not materially differ from Anderson. The following quotation from Henry's History of Great Britain, giving the particulars of Octher's narrative, will not be uninteresting: "In addition to all these [expeditions against the Danish pirates] and many other cares, Alfred encouraged foreigners that were in his service, and some of his own subjects, to undertake voyages for making discoveries, and opening new sources of trade, both toward the north and south; of which it will be proper to give some account. There is still extant a very curious relation of one of these voyages, undertaken by one Ochter, a Norwegian. This relation was given by the adventurer himself at his return, and written down from his mouth by King Alfred with his own hand. The style of this precious fragment of antiquity, is remarkably simple, and it seems to have been designed only as a memorandum for the king's own private use. This simplicity of style is imitated in the following translation from the original Saxon, of that part of it which it is thought necessary to lay before the reader. Ochter, after giving a very curious description of the country inhabited by the Finmans, proceeds to say, that 'he visited this country also with a view of catching horse-whales, which had bones of very great value for their teeth, of which he brought some to the king; that their skins were good for making ropes for ships. These whales are much less than other whales, being only five ells long. The best whales were catched in his own country, of which some were forty-eight, some fifty yards long. He said that he was one of six who had killed sixty in two days.'"

These horse-whales, spoken of by Ochter, were what we call sea-horses, and the Dutch sea-cows, or morses. It is probable the length of the whales caught in his own country is greatly exaggerated. Beal quotes from many of the ancient writers instances of extraordinary exaggerations of this kind, and doubts whether any whales were ever seen of a greater length than eighty or ninety feet, even admitting that they were once found of a larger growth than any now seen or captured.

The earliest authentic data that I have been able to find respecting the origin of the whale fishery, as a regular and permanent branch of trade, is that furnished by M'Culloch, in his Commercial Dictionary, which, although little more than a condensation from the works of Anderson, Macpherson, and others, is of a more reliable character than any similar compilation I have met with. It has often been asserted that the whale fishery had its origin on the North American coast, and that the inhabitants of Plymouth province were the first to engage in it as a systematic business. We can justly claim the honor of having done more to extend and render profitable this branch of trade, under very discouraging circumstances, than any nation upon earth, but not that of having commenced it before other nations. At the time the inhabitants of Massachusetts were making their first attempts in the capture of the whale (about 1650), the Biscayans had already extensively engaged in that business, the Dutch and the English had followed their example, the Russia Company had obtained an exclusive charter for it, and many other nations of Europe had directed their attention to the northern fisheries.

"It is probably true, as has been sometimes contended," says M'Culloch, "that the Norwegians occasionally captured the whale before any other European nation engaged in so perilous an enterprise. But the early efforts of the Norwegians were not conducted on any systematic plan, and should be regarded only in the same point of view as the fishing expeditions of the Esquimaux. The Biscayans were certainly the first people who prosecuted the whale fishery as a regular commercial pursuit. They carried it on with great vigor and success in the 12th, 13th, and 14th centuries. In 1261, a tithe was laid upon the tongues of whales imported into Bayonne, they being there a highly-esteemed species of food. In 1388, Edward III. relinquished to Peter de Puayanne a duty of £6 sterling a whale, laid on those brought into the port of Biaritz, to indemnify him for the extraordinary expenses he had incurred in fitting out a fleet for the service of his majesty. This fact proves beyond dispute that the fishery carried on from Biaritz at the period referred to must have been very considerable indeed; and it was also prosecuted to a great extent from Cibourre, Vieux Boucan, and, subsequently, from Rochelle and other places.

"The whales captured by the Biscayans were not so large as those that are taken in the Polar Seas, and are supposed to have been attracted southward in the pursuit of herrings. They were not very productive of oil, but their flesh was used as an article of food, and the whalebone was applied to a variety of useful purposes, and brought a very high price.

"This branch of industry ceased long since, and from the same cause that has occasioned the cessation of the whale fishery in many other places—the want of fish. Whether it were that the whales, from a sense of the dangers to which they exposed themselves in coming southward, no longer left the Icy Sea, or that the breed had been nearly destroyed, certain it is that they gradually became less numerous In the Bay of Biscay, and at length ceased almost entirely to frequent that sea; and the fishers being obliged to pursue their prey upon the banks of Newfoundland and the coasts of Iceland, the French fishery rapidly fell off.

"The voyages of the Dutch and English to the Northern Ocean, in order, if possible, to discover a passage through it to India, though they failed of their main object, laid open the haunts of the whale. The companions of Barentz, who discovered Spitzbergen in 1596, and of Hudson, who soon after explored the same seas, represented to their countrymen the amazing number of whales with which they were crowded. Vessels were in consequence fitted out for the Northern whale fishery by the English and Dutch, the harpooners and part of the crew being Biscayans. They did not, however, confine their efforts to a fair competition with each other as fishers. The Muscovy Company obtained a royal charter, prohibiting the ships of all other nations from fishing in the seas round Spitzbergen, on pretext of its having first been discovered by Sir Hugh Willoughby. There can, however, be no doubt that Barentz, and not Sir Hugh, was its original discoverer; though, supposing that the fact had been otherwise, the attempt to exclude other nations from the surrounding seas, on such a ground, was not one that could be tolerated. The Dutch, who were at that time prompt to embark in any commercial pursuit that gave any hopes of success, eagerly entered on this new career, and sent out ships fitted equally for the purposes of fishing, and of defense against the attacks of others. The Muscovy Company having attempted to vindicate its pretensions by force, several encounters took place between their ships and those of the Dutch. The conviction at length became general that there was room enough for all parties in the Northern Seas; and in order to avoid the chance of coming into collision with each other, they parcelled Spitzbergen and the adjacent ocean in districts, which were respectively assigned to the English, Dutch, Hamburghers [*sic*], French, Danes, &c.

"The Dutch, thus left to prosecute the fishery without having their attention diverted by hostile attacks, speedily acquired a decided superiority over all their competitors.

"When the Europeans first began to prosecute the fishery on the coast of Spitzbergen, whales were every where found in vast numbers. Ignorant of the strength and stratagems of the formidable foe by whom they were now assailed, instead of betraying any symptoms of fear, they surrounded the ships and crowded all the bays: Their capture was, in consequence, a comparatively easy task, and many were killed which it was afterward necessary to abandon, from the ships being already full.

"While fish were thus easily obtained, it was the practice to boil the blubber on shore in the North, and to fetch home only the oil and whalebone; and perhaps nothing can give a more vivid idea of the extent and importance of the Dutch fishery in the middle of the 17th century, than the fact that they constructed a considerable village, the houses of which were all previously prepared in Holland, on the Isle of Amsterdam, on the northern shore of Spitzbergen, to which they gave the appropriate name of *Smeerenberg* (from *smeeren*, to melt, and *berg*, a mountain). This was the grand rendezvous of the Dutch whale ships, and was amply provided with boilers, tanks, and every sort of apparatus required for preparing the oil and bone. But this was not all. The whale ships were attended with a number of provision ships, the cargoes of which were landed at Smeerenberg, which abounded during the busy season with well-furnished shops, good inns, &c.; so that many of the conveniences and enjoyments of Amsterdam were found within about eleven degrees of the pole! It is particularly mentioned that the sailors and others were every morning supplied with what a Dutchman regards as a very great luxury, *hot rolls* for breakfast. Batavia and Smeerenberg were founded nearly at the same period, and it was for a considerable time doubted whether the latter was not the more important establishment.—(*De Reste, Histoire des Peuches*, &c., tome i., p. 42.)

"During the flourishing period of the Dutch fishery, the quantity of oil made in the North was so great that it could not be carried on by the whale ships; and every year vessels were sent out in ballast to assist in importing the produce of the fishery.

"But the same cause which had destroyed the fishery of the Biscayans, ruined that which was carried on in the immediate neighborhood of Spitzbergen. Whales became gradually less common, and more and more difficult to catch. They retreated first to the open seas, and then to the great banks of ice on the eastern coast of Greenland. When the site of the fishery had been thus removed to a very great distance from Spitzbergen, it was found most economical to send the blubber direct to Holland. Smeerenberg was, in consequence, totally deserted, and its position is now with difficulty discernible."

"When in the most flourishing state, toward 1680, the Dutch whale fishery employed about 260 ships and 14,000 sailors."

This brings us nearly up to the period when the settlers in North America began to fit out vessels for this service. It will be seen, however, that they had been for some time previously engaged in the fishery near the shores, in boats and canoes.

My attention was directed by the Hon. Mr. Grinnell, a gentleman better acquainted with the history of the whale fishery than perhaps any man in the United States, to a very curious and entertaining account of the rise and progress of the whale fishery, by an old resident of Nantucket, Mr. Obed Macy,[1] a descendant of Thomas Macy, the first settler. From this work I glean the following facts in relation to the first at-tempts made by the inhabitants of Nantucket to capture the "great leviathan of the deep."

Driven by religious persecution from his home in Massachusetts, whither he had emigrated from England in 1640, Thomas Macy removed with his family to the Island of Nantucket in the fall of 1659, where he established a settlement.

The whaling business was not commenced till several years after the settlement of the island. In the interval, the people were occupied in farming, or in fishing near the shores. Fish were plenty, and easily caught. The Indians were instructed in the mode of fishing practiced by the whites, and, in return, the whites were assisted by the Indians in pursuing the business. Previous to their acquaintance with the English, the natives fished with a rude line of twisted grass, to which they attached a stone for a sinker, and a clumsy hook of bone. Some of the sinkers just mentioned remain to this day. They resemble a ship's block in form, and weigh two or three pounds. With this inconvenient apparatus, they caught but few fish compared with the number obtained by the better-adapted hook and line of the Europeans.

"The first whaling expedition was undertaken by some of the original purchasers of the island, the circumstances of which are handed down by tradition. A whale, of the kind called 'scragg,' came into the harbor, and continued there three days. This excited the curiosity of the people, and led them to devise measures to prevent his return out of the harbor. They accordingly invented, and caused to be wrought for them, a harpoon, with which they attacked and killed the whale. This first success encouraged them to undertake whaling as a permanent business, whales being at that time numerous in the vicinity of the shores. In furtherance of their design, they made a contract with James Lopar to settle on the island and engage in the business."

Lopar made a contract with some of the islanders, and a company was organized for the purpose of carrying on the business systematically. An agreement was drawn up and duly signed, allowing Lopar one third of the proceeds, and two thirds to those who embarked in the enterprise with him.

"As it now appeared that there was a prospect of carrying the business of whaling into effect, the town, willing to give it every encouragement that it required, agreed with John Savage to remove thither with his family, and to serve them in the occupation of cooper; and, to induce his compliance, to give him ten acres of land, and commonage for three cows and one horse—nearly on the same conditions as those above mentioned in relation to Lopar.

"How far this plan succeeded we are in a great measure unacquainted; the profits of the business, however, were sufficient to encourage its pursuit. Finding that the people of Cape Cod had made greater proficiency in the art of whale catching than themselves, the inhabitants, in 1690, sent thither and employed a man by the name of Ichabod Paddock, to instruct them in the best manner of killing whales and extracting their oil. The pursuit of whales commenced in boats from the shore, and increased from year to year till it became the principal branch of business with the islanders. The Indians, ever manifesting a disposition for fishing of every kind, readily joined with the whites in this new pursuit, and willingly submitted to any station assigned them. By their assistance the whites were enabled to fit out and man a far greater number of boats than they could have done of themselves. Nearly every boat was manned in part, many almost entirely, by natives: some of the most active of them were made steersmen, and some were allowed now to head the boats: thus encouraged, they soon became experienced whalemen, and capable of conducting any part of the business.

"They sometimes, in pleasant days during the winter season, ventured off in their boats nearly out of sight of land. It has often been remarked by the aged, that the winters were not so windy and boisterous at that time as at present, though quite as cold; and that it would sometimes continue calm a week, or even a fortnight.

"The process called saving the whales after they had been killed and towed ashore, was to use a crab, an instrument similar to a capstan, to heave and turn the blubber off as fast as it was cut. The blubber was then put into their carts and carried to their try-houses, which, at that early period, were placed near thee dwelling-houses, where the oil was boiled out and fitted for market.

"To enable them to discover whales at a considerable distance from the land, a large spar was erected, and cleats fixed to it, by which the whalemen could climb to the top, and there keep a good look-out for their game. There was no perceptible decrease of the number of whales during the period of the first thirty or forty years from the commencement of the fishery. It appears that in 1726 they were very numerous, for eighty-six were taken in that year, a greater number than were obtained in any one year either before or since that date. The greatest number ever killed and brought to shore in one day was eleven. This mode of whaling continued until about the year 1760, when the whales became scarce, and it was by degrees discontinued. Since that date, whales have only occasionally been obtained by boats from the shore.

"It is remarkable that, notwithstanding the people had to learn the business of whaling, and to carry it on under many hazardous circumstances, yet not a single white person was killed or drowned in the pursuit, in the course of seventy years preceding 1760. The whales hitherto caught near the shores were of the right species.

"The first spermaceti whale known to the inhabitants was found dead, and ashore, on the southwest part of the island. It caused considerable excitement, some demanding a part of the prize under one pretence, some under another, and all were anxious to behold so strange an animal. There were so many claimants of the prize, that it was difficult to tell to whom it should belong. The natives claimed the whale because they found it; the whites, to whom the natives made known their discovery, claimed it by a right comprehended, as they affirmed, in the purchase of the island by the original patent. An officer of the crown made his claim, and pretended to seize the fish in the name of his majesty as being property without any particular owner. After considerable discussion between these contending parties, it was finally settled that the white inhabitants who first found the whale should share the prize equally among themselves. The teeth, which were considered very valuable, had been extracted by a white man and an Indian before any others had any knowledge of the whale. All difficulty being now settled, a company was formed, who commenced cutting the whale in pieces convenient for transportation to their try-works. The sperm produced from the head was thought to be of great value for medical purposes. It was used both as an internal and an external application and such was the credulity of the people, that they considered it a certain cure for all diseases; it was sought with avidity, and, for a while, was esteemed to be worth its weight in silver. The whole quantity of oil obtained from this whale is not known.

"Whales being plenty near the shores, people were led to conclude that they should find them still more numerous were they to pursue them with vessels into the deep. That the pursuit of whales into the ocean was early anticipated, we know by an anecdote related by one of our ancestors. In the year 1690, the same in which Ichabod Paddock was sent for from Cape Cod, as before related, some persons were on a high hill, afterward called Folly House Hill, observing the whales spouting and sporting with each other, when one observed, '*There,*' pointing to the sea, '*is a green pasture where our children's grand-children will go for bread.*' It was many years, however, before they began to whale with vessels, but at what precise time it happened we have no means of knowing.

"Previous to whaling in vessels, it was necessary to determine where the harbor should be. It has already been mentioned that the one at Madacket was at first preferred; but this was afterward relinquished for the present harbor, which is larger, more land-locked, and in many other respects better adapted to the purposes of navigation than the first-mentioned. The south side of this harbor was first selected for the site of the town; the proprietors, therefore, laid out house lots, or homesteads, of one hundred rods in length, and three or four rods in width; but many inconveniences were afterward found to attend this location, and the present situation of the town was soon after selected. It being now determined where the town should be, it became necessary to give it a name, and it was accordingly called *Sherburne*, by order of Francis Lovelace, Esq., governor of the province of New York, in his written directions bearing date April 18th, 1673.

"The first spermaceti whale taken by the Nantucket whalers was killed by Christopher Hussey. He was cruising near the shore for right whales, and was blown off some distance from the land by a strong northerly wind, where be fell in with a school of that species of whales, and killed one and brought it home. At what date this adventure took place is not fully ascertained, but it is supposed to be not far from 1712. This event gave new life to the business, for they immediately began with vessels of about thirty tons to whale out in the 'deep,' as it was then called; to distinguish it from shore whaling. They fitted out for cruises of about six weeks, carried a few hogsheads, enough, perhaps, to carry the blubbers of one whale, with which, after obtaining it, they returned home. The owners then took charge of the blubber, and tried out the oil, and immediately sent the vessel out again. At the commencement of this mode of whaling, it was found necessary to erect try-houses near the landing, and a number were built on the beach a little south of the wharves. North from these they erected small buildings, called warehouses, in which they put their whaling apparatus and other outfits.

"In 1715, the number of vessels engaged in the whaling business was six, all sloops, from thirty to forty tons burden each, which produced £1100 sterling, or $4888 88 cents.

"As the shipping increased, it was found indispensably necessary to have wharves. The first built is that now called straight Wharf, constructed in 1723. Previous to this, there had been places built off, called landing-places, which were but temporary, and were often broken up by winter storms.

"The island was now in a flourishing condition. The inhabitants were fast increasing in number and wealth; the land was principally purchased of the natives; it was very productive when improved; the natives were cordially enlisted in the service of the whites; fish and fowls were plenty; the whaling had become a most profitable employment, and had promised business for all. What a prospect must this have been to a people like them, remarkable for their industry and prudence, never so well pleased as when they had as much business as they could perform. This being the general character of the inhabitants, they increased in wealth as fast as could be expected. This business, it is true, did not afford great profits—less, perhaps, than almost any other; but the people, being of a frugal disposition, required but little to keep them comfortable, and there were but few among them who aspired after great things.

"As the whaling business was found to answer their expectations, they were encouraged to increase the number and size of their vessels. Sloops and schooners of from forty to fifty tons were put into the business. Vessels of this size being supposed to be best adapted to whaling near the coast, no larger ones were employed for many years. At length whales began to be scarce near the shore, and some enterprising persons procured larger vessels and sent them out to the southward, as it was called, where they cruised until about the first of the seventh month, when they came in and refitted, and went to the eastward of the Grand Bank, where they continued through the whaling season, unless they completed their lading sooner, which frequently happened. The vessels that went on these voyages were generally sloops of sixty or seventy tons; their crews were made up in part of Indians, there being usually from four to eight in each vessel. They were pleased with the business, and always ready to engage in it when called upon.

"At the close of the whaling season, the vessels were mostly drawn on shore for the winter, being considered safer and less expensive in that situation than at the wharves. The boats were placed on the beach bottom upward, and tied together, to prevent disasters in gales of wind, and all the whaling gear was put into the warehouses."

Having followed Mr. Macy thus far in his very interesting account of the origin of the whale fishery, I shall endeavor to preserve a sort of connection of the subject by introducing an outline of the rise and progress of this pursuit at other points of the American Continent, and in Europe. It does not appear that Mr. Macy claims for the early settlers of Nantucket the honor of having been the first people near the shores of this Continent to engage in the capture of the whale. This is a matter about which historians differ, and I believe it is conceded by all to be involved in much obscurity. The islanders, it appears, were encouraged, by their success in capturing a whale which had come into the harbor, to make a contract with one James Lopar, "to settle on the island and engage in the business." From this it is but reasonable to infer that Loper was not a resident of the island, and that he must have had some experience elsewhere in the catching of whales. Next it appears that the people of Cape Cod, having made "greater proficiency in the art of whale catching than themselves, the inhabitants in 1690 sent thither and employed a man by the name of Ichabod Paddock, to instruct them in the best manner of killing whales and extracting their oil." How long the people of Cape Cod had been engaged in the business, does not appear; and, indeed, after a careful investigation of the subject, I am constrained to believe this interesting point has never been recorded, and can not now be ascertained, otherwise than by tradition. The precise date at which the first whale was captured on the island of Nantucket is not known, though it appears that this event took place some time after the settlement by the original purchasers in 1660. Mr. Richard Norwood, who resided at the Bermudas, states, as will be seen from the extracts which I am about to introduce, that the whale fishery had been carried on in the bays of those islands for "two or three years" prior to 1667. Putting these statements together—"several years after the settlement of the island" (Nantucket, in 1660), and "two or three years" prior to 1667, at the Bermudas Islands, with the proficiency of the people of Cape Cod in 1690, and it is very difficult to decide to whom the honor is due. I am greatly aided in 'my researches upon this point by the following compilation from the work of a distinguished English writer, who seems to have devoted more than ordinary attention to the history of the sperm whale. Having given a sketch of the rise and progress of the whale fishery on the island of Nantucket, it will be interesting to glance for a moment at the earliest records extant, touching the same subject, elsewhere. In the "Natural History of the Sperm Whale," a valuable scientific work by Thomas Beale, Esq, surgeon to the "Kent" and "Sarah Elizabeth," London whalers, and Demonstrator of Anatomy to the Eclectic Society, I find the following passages in reference to the early history of the whale fishery, and its progress in Europe. It will be seen that Dr. Beale gives us the credit of having been the first to engage in this business, though he claims for the mother country the honor of having been the first to open the sperm fishery in the Pacific.

"The origin of the sperm whale fishery—that is, before it became organized as a branch of commerce—like the origin of other fisheries of the same nature, is involved in such deep mystery as almost altogether to defy the searching acumen of the historian. Without looking into the ancient, romancing, and classical histories with which most of the countries of Europe abound, and which contain wonderful stories of the appearance, death, or capture of the sperm whale, or other creatures of the same order, it may be sufficient for some of us to know, that during the early part of the last century, a few daring individuals, who inhabited the shores of the American Continent, fitted out their little crafts, furnished with weak and almost impotent weapons, to attack and destroy in its own element the mighty monarch of the ocean, in order to rob his immense carcass of the valuable commodity with which it is surrounded. But even as far back as the year 1667, we find a letter, published in the second volume of the *Philosophical Transactions,* from Mr. Richard Norwood, who resided at the Bermudas, which states that the whale fishery had been carried on in the bays of those islands for 'two or three years,' evidently meaning the black whale fishery; for in other parts he says, 'I hear not that they have found any spermaceti in any of these whales;' but subsequently he states in the same letter, 'I have heard from credible persons that there is a kind of whale having *great teeth*, as have the spermaceti at Eleuthera, and others of the Bahama Islands. One of this place,' he continues, 'John Perinchief, found one there dead, driven upon an island, and though, I think, ignorant in the business, yet got a great quantity of spermaceti out of it.'

"He says again: 'It seems they have not so much oil as ours (meaning the black whale), but the oil, I hear, is at first like spermaceti, but they clarify it, I think, by the fire.' But in vol. iii., *Phil. Trans.*, in a letter from the same place, written a year or two afterward, we find something like a beginning of the sperm whale fishery threatened by a Mr. Richard Stafford, who informs us that he has killed several black whales himself, and who is represented as a very intelligent gentleman. He says, 'Great stores of whales make use of our coast;' but in another part he states, 'but here have been seen spermaceti whales driven upon the shore; these have divers teeth, about the bigness of a man's wrist. I have been,' says he, 'at the Bahama Islands, and there have seen this same sort of whale dead on the shore, with sperma all over their bodies. *Myself, and about twenty others, have agreed to try whether we can master and kill them, for I could never hear of any of that sort that was killed by any man, such is their fierceness and swiftness.*' He concludes by remarking, that 'one such whale would be worth many hundred pounds." A weighty reason for the establishment of the fishery, no doubt. The same writer, in another part of his letter, states, 'There is one island among the Bahamas, which some of our people are settled upon, and more are going thither. It is called *New Providence,* where many rare things might be discovered, if the people were but encouraged.' This same *New Providence* afterward became so famous as a whale-fishing station, by the exertions of our American descendants.

"But even before these needy adventurers commenced their career of spermaceti hunting, we have had it proved to us that the Indians who inhabited the shores of America used to voyage out to sea and attack this animal from their canoes, and pierce him with their lances of wood, or other instruments of the same material, which were barbed, and which, before they were plunged into his flesh, were fastened by a short warp, or piece of rope, to a large block of light wood, which was thrown overboard the moment the barbed instrument was thrust into its body, which, being repeated at every rising of the whale, or when they were so fortunate as to get near enough to do so, in a few instances, by a sort of worrying-to-death system, rewarded the enterprising savage with the lifeless body of his victim, but which, in most cases, was that of a very young one; and even this, when towed to the shore, it was impossible for them to turn over, so that they were obliged to content themselves with flinching the fat from one side of the body only. Few, indeed, must these instances have been, when we consider the means that were employed in the capture of so immense an animal, possessing such enormous strength, by which their barbed spears or lances of wood must have been frequently shivered to atoms, or drawn from the flesh of the whale, by the resistance the blocks of wood to which they were attached must have occasioned when the animal became frightened into its utmost speed; and when we know at the present time that by their powerful actions and convulsive movements the best tempered iron, of which our harpoons and lances are made, frequently becomes twisted to pieces, while the boats which are used in the chase are often thrown high into the air with the head, or broken to fragments by one blow of the tail of this enormous creature.

"But although, as has been before stated, Mr. Richard Stafford had threatened to commence the sperm whale fishery at the Bahama Islands, it appears rather doubtful whether he did so or not, when we come to peruse the letter of the Hon Paul Dudley, F.R.S., published in 1724 (*Phil. Trans.*, vol. xxxiii.), an extract of which states, 'I very lately received of one Mr. Atkins, an inhabitant of *Boston*, in *New England*, who used the Whale fishery for ten or twelve years (black whales), and was *one of the first* that went out a fishing for the *spermaceti whales*, about the year 1720.' It also appears in this account that the fishery even then was very little understood, for Mr. Atkins himself says, 'He never saw, nor certainly heard of a spermaceti female being taken in his life;' for he states, 'the cows of that species of whale being much more timorous than the males, and almost impossible to come at, unless when haply found asleep upon the water, or detained by their calves.'

"In another part of his letter the Hon. Paul Dudley states: 'Our people formerly used to kill the whale near the shore, but now they go off to sea in *sloops* and *whale boats*. Sometimes,' he says, 'the whale-is killed by a single stroke, and yet at other times she will hold the whalemen in play *near half a day together* with their lances, and sometimes they will get away after they have been lanced, and spouted thick blood, with irons in them and drags (drouges) fastened to them, which are thick boards about fourteen inches square.'

"But even after the capture of the sperm whale had occasionally been carried on in ships by the descendants of the European settlers upon the American shores, who struck the whales with the harpoon, having a log of wood attached, after the Indian fashion, it was a considerable time before any great improvement manifested itself in their mode of fishing. Presumptuous indeed was he deemed who first proposed to chase and capture such huge beings in small boats, and by the aid of lines at the end of which was attached the harpoon, by which they could draw themselves to the harpooned whale whenever they wished to destroy it with the lance.

"An American whaler, who had been bred from his boyhood in the service, informed me that his grand-father had been employed on a whaling expedition in a small vessel off the coast of America, and that, having experienced a great deal of ill success in consequence of their being unable to capture any whales by means of the log harpoon, the captain of their little barque wished them to make trial of the method of which they had just heard, by the boat and line; but to his irresolute seamen the idea seemed monstrous; the mere thought of having the boat they were in attached to an infuriated leviathan by a strong rope struck terror among the whole crew. 'What,' said they, 'shall we be dragged to the bottom of the sea? Shall we be towed with the velocity of lightning to the other side of the world? Shall we be torn to pieces by the jaws of the monstrous fish that we may be fastened to?' In vain did their captain explain to them the various means they could employ to avoid those anticipated dangers; he urged their reason to note the excellence of the plan, but his eloquence proved of no avail; so fearful were they of this dangerous innovation on their old method, that the very rope which the captain had prepared for the service was pointed through the ship's stern during the night, and allowed to run overboard. But, nevertheless, others more daring undertook the trial soon afterward, in which they frequently came off victorious, so that the new method was established among them, and has since been much improved.

"The fishery was thus carried on at first by a few individuals in America from their own shores; but, as their numbers increased, the quantity of whales diminished; so that, in a few years, they had not only destroyed great numbers of these useful animals, but had driven the remainder to find more secure retreats, in which they could follow their natural inclinations without being harassed by the chase or wounded by the harpoon.

"But, about the year 1771, we find that the American navigators were engaged with extraordinary ardor in the whale fisheries which were carried on in the North and South Atlantic Oceans. From the year 1771 to 1776, Massachusetts alone employed annually 183 vessels, carrying 13,820 tons in the former, and 121 vessels, carrying 14,026 tons in the latter.

"Mr. Burke, in his famous speech on American affairs in 1774, adverted to this wonderful display of daring enterprise in the following eloquent words: 'As to the wealth,' said he, 'which the colonists have drawn from the sea by their fisheries, you had all that matter fully opened at your bar. You surely thought these acquisitions of value, for they seemed to excite your envy; and yet the spirit by which that enterprising employment has been exercised, ought rather, in my opinion, to have raised your esteem and admiration. And pray, sir, what in the world is equal to it? Pass by the other parts, and look at the manner in which the New England people carry on the whale fishery. While we follow them among the trembling mountains of ice, and behold them penetrating into the deepest frozen recesses of Hudson's and Davis's straits—while we are looking for them beneath the Arctic circle, we hear that they have pierced into the opposite region of the polar cold—that they are at the antipodes, and engaged under the frozen serpent of the south. Falkland Island, which seems too remote for the grasp of national ambition, is but a stage and resting-place for their victorious industry. Nor is the equinoctial heat more discouraging to them than the accumulated winter of both poles. We learn that, while some of them draw the line or strike the harpoon on the coast of Africa, others run the longitude, and pursue their gigantic game along the coast Of Brazil. No sea but what is vexed with their fisheries, no climate that is not witness of their toils. Neither the perseverance of Holland, nor the activity of France, nor the dexterous and firm sagacity of English enterprise, ever carried this most perilous mode of hardy industry to the extent to which it has been pursued by these recent people—a people who are still in the gristle, and not hardened into manhood.'

"Whether this eloquent address had any effect or not upon the minds of our own merchants and ship-owners in stimulating them to fit out ships for the sperm and other whale fisheries, I am not aware, but it is certain that in the following year (1775) the first attempt was made to establish the sperm whale fishery from Britain; and we accordingly find, from private statements on which I can securely rely, that ships of from 100 to 109 tons burden were sent to South Greenland, the coast of Brazil, the Falkland Islands, and the Gulf of Guinea, for the purpose of procuring sperm and other oils. The names of the ships which were thus employed in these distinct expeditions were the 'Union,' 'Neptune,' 'Rockingham,' 'America, 'Abigail,' 'Hanover,' 'Industry,' 'Dennis,' 'Beaver,' and 'Sparrow;' but the principal places of resort of the spermaceti whale not having been yet discovered, these vessels met with very trifling success.

"In the following year, 1776, the government, with a view to stimulate all persons engaged in these fisheries, established a principle of reward for those ships which were most successful in their endeavors; in accordance with which, five different bounties or premiums were offered, forming a scale of prizes for those who were so fortunate as to prove the five gradations of success; the sum of five hundred pounds being the maximum, and that of one hundred being the minimum prize.

"In 1781, four ships were fitted out for the River St. Lawrence, but after they had been out a considerable time, they returned with the discouraging announcement of having only procured six gallons of sperm oil among them during the whole time of their absence."

Note

1. The History of Nantucket: being a compendious Account of the first Settlement of the Island by the English, together with the Rise and Progress of the Whale Fishery, and other historical Facts relative to said Island and its Inhabitants. By Obed Macy. 1835.

From Charles Wilkes, "Currents and Whaling," *Narrative of the United States Exploring Expedition*

Vol. 5 (Philadelphia: C. Sherman, 1844)

INTRODUCTION

Almost as soon as the War of 1812 ended, there was a movement to expand American trade abroad. Finding new markets and new resources required an investment the young republic was reluctant to make, and it was not until the mid-1830s that it appeared that an exploring expedition would be sponsored by the government. When appropriations were finally made in response to the lobbying efforts of enthusiasts like Jeremiah N. Reynolds, the enterprise was widely viewed in the Navy as a career-ending assignment, and Wilkes (1798–1877), only a lieutenant at the time, was given command of the United States Exploring Expedition in 1838 after a number of senior officers turned it down. He was an excellent geographer but a tactless leader, and much of his achievement was lost amid controversy surrounding the expedition's return in 1842. Melville purchased his set of the *Narrative* in April 1847. Sealts #532; Bercaw #759. Modern edition: Upper Saddle River: Gregg Press, 1970.

PRIMARY SOURCE

It is well known to whalers that the favourite and appropriate food of the sperm whale is a gelatinous medusa; which, however, has not as yet received from naturalists much attention. It may, however, be advanced as certain that this molluscous animal most abounds in the higher latitudes of both hemispheres, which would therefore seem to be the places in which it is produced, and to which its habits are best adapted.[1] During our cruise in the higher southern latitudes, we saw vast numbers of these medusæ around and near the icebergs. The quantity was such as to prove conclusively, that it was in the waters of the temperature caused by the vicinity of these masses of ice, that they delight to dwell. Whales were also in abundance, and although principally of the fin-back species, sperm whales were not entirely wanting.

As regards the medusa, its powers of locomotion are feeble, and confined chiefly to the purpose of rising and sinking at pleasure. If polar currents exist, it must therefore be swept by them from the place of its nativity, and in its passage to lower latitudes, will by its locomotive power seek strata in the water of the low temperature to which its constitution is best adapted. My attention was drawn to the habits of the whales here in particular, from the novel manner they exhibited of feeding near the surface instead of diving lower down, as they are usually seen to do in lower latitudes: they were constantly in sight, instead of being only seen at intervals.

It will be readily admitted that the medusa, like other animals, has its appropriate seasons of procreation, and it will appear probable that the season at which we saw them in such numbers was that in which they are brought forth most abundantly. So also, however low the temperature of the water in which they delight, there is little probability that their increase goes forward when the regions in which we met them are locked up in ice, and the genial light and warmth of the sun is denied them.

The food of the sperm whale will therefore be borne off to lower latitudes by the polar streams in greater abundance at one season than another, and this former season corresponds with that in which these currents have their greatest force. The sperm whale, it must be expected, will leave the higher latitudes and follow the currents which transport his food.

In conformity with this view, we find the habits of the sperm whale migratory. The polar currents, as has been seen, disappear from the surface in many cases, but do not cease to flow; and even when felt both at the surface and below, they will in approaching lower latitudes have their higher temperatures near the surface. The medusa will therefore descend in either case to greater depths, and the whale must dive in quest of the food which in higher latitudes he could find at the surface. We have seen in what a decided manner the polar currents become or continue superficial at the southern promontories of the continents. A similar cause operating to a less extent raises them if submarine, when they are interrupted or impeded by islands, and spreads water of low temperature over the surface. Here then, at the proper season, the food of the whale will be not only more accessible, but more abundant within a given space, in consequence of the check the velocity of the stream must experience.

So also in the zones of calms we have seen that the matter borne by the polar currents in all probability finds a resting-place; and here also, at fit seasons, the food of the whale must be abundant. Points possessing either of these characteristics I have distinguished, as before stated, by the name of nuclei.

However satisfactory this theory may be in explaining the causes of the migratory habits of the sperm whale, it is obvious that we do not know enough of the natural history of his favourite food, nor of the rate and course of all the submarine polar currents, to enable us to predict with certainty the seasons at which he will be found in particular parts of the ocean. This can be learned by observation alone, and long experience has taught those who are skillful in the whale-fishery the position of the favourite haunts of their prey, and the times at which they are most likely to be met with there. Comparing these points and the nuclei of the currents, as observed and explained in the preceding pages, the coincidence will strike every one [*sic*] who will examine the subject; and when all the facts necessary to illustrate this subject shall be ascertained, theory may serve in some degree to shorten the apprenticeship which is now necessary in order to acquire the requisite knowledge of the places and seasons wherein to meet the game in its adventurous employment; the object therefore of the residue of this chapter will be devoted to whaling, and to point out the results which our own observations, with the information derived from others, has afforded.

The whaling interest, taking into consideration the extent to which it has been carried by our countrymen, may be almost claimed as peculiarly American. There are few employments in which the enterprise and industry of our countrymen are so well developed as in this, or in which so much hardihood or so many resources are required to insure success.

Our whaling fleet may be said at this very day to whiten the Pacific Ocean with its canvass, and the proceeds of this fishery give comfort and happiness to many thousands of our citizens. The ramifications of the business extend to all branches of trade, are spread through the whole Union, and its direct or secondary influence would seem to recommend it to the especial protection and fostering care of the government.

As it was among the first objects of the Exploring Expedition to render the dangerous path of these enterprising mariners more safe, I trust it will have been perceived, that throughout the operations of the squadron, this interest has never been lost sight of. In fact, it has always been my constant study to endeavour to accomplish whatever could tend to its benefit. In the course of the various and devious voyages we have made, the greatest attention has been paid to the winds and currents; and from my investigations, I hope to be able to point out the most feasible routes by which to gain the proper cruising-grounds, and to define their localities more clearly than has hitherto been done.

Note

1. Innumerable animalculæ, the appropriate food of the right whale, are also found there, as has been seen by our own observations at the south, and those of Scoresby at the north.

From Pierre Bayle, *An Historical and Critical Dictionary*

Vol. 3 (London: Harper, Brown, *et al.*, 1710)

INTRODUCTION

Bayle (1647–1706) was raised a French Calvinist. He briefly converted to Roman Catholicism before settling in Geneva and readopting Calvinism. He returned to France and pursued a checkered career as a professor of philosophy, suffering persecution and harassment, and he essentially died a religious exile in the Netherlands. An Enlightenment philosopher who is sometimes credited with inventing the encyclopedia (of which this entry is a sample) and a skeptical Protestant, Bayle surrounded each entry with copious marginalia (mostly citations) and footnotes (explication). The "Jonas" entry was surrounded with a further thousand words of annotation. Melville obtained his set before April 1849. See Millicent Bell, "Pierre Bayle and *Moby-Dick*," *PMLA* 66:5 (September 1951), 626–648. Sealts #51; Bercaw #50.

PRIMARY SOURCE

JONAS, One of the *Jewish* Prophets. Since most of the things relating to him may be found in two other Dictionaries, I will only insist upon a few Particulars. Some *Rabbins* have been so silly as to say, that being first swallow'd by a Male Fish, he was afterwards Discharged into the Body of a Female. Not finding himself press'd in his first Prison, say they, he had no

Recourse to the Invocation of God; upon which the Male Fish had Orders to Disgorge him into the Stomach of a Female, which was Big. He there perceiv'd himself in too strait [*sic*] a Lodging, and pronounc'd the noble Canticle, which he has left us, and which appeas'd the Wrath of Heaven. They who Confute this Story with this Argument, that a Whale that was Impregnated would not have press'd *Jonas*, unless he had Lodg'd in her Womb, make but a weak Objection. We have seen in another Place that the Heathen Poets have given out an Event of their *Hercules* somewhat resembling this. They stole it from the Sacred Story, and Falsified it, according to their Fancy. This is at least the common opinion of our Authors. The Antient Fathers [*sic*] thought it strange that the Heathens should Reject this History of *Jonas*, after having Adopted the Fable of *Hercules*. Those who say that this Prophet came out of the Fish's Belly in the Port of *Ninive*, are but little skilled in Geography; nor is it probable that he came out of it upon the Coast of the *Euxine* or of the Red Sea. 'Tis much more credible he was cast upon the dry Land near the City of *Joppe* where he took Shipping. Some confirm this Conjecture by the Fable of *Andromeda*; for they pretend that *Jonas's* Adventure laid a Foundation for the Poetical Narrations concerning *Andromeda's* being exposed to the Fury of a Sea-Monster, which happened near *Joppe*. . . . Here is another idle story of the *Rabbins*. They say that the Fish which swallowed *Jonas* had seven Eyes, which served as so many Windows for the Prophet to see all that was in the Sea, and amongst other things, the way which the *Israelites* took in Crossing the Red Sea. Those who take him for the Young Man who was sent by *Elisha* to *Jehu*, to give him the Royal Unction, deserve not to be heard. In St. *Jerome's* time the Tomb of *Jonas* was still shewn in the place of his Birth. Mr. *Simon* Affirms that the *Turks have Built a very fine Mosque to the Honour of* Jonas, *in which there is a Miraculous Lamp, that Burns continually without any Oil or other Liquor, if we will give credit to their Fooleries*. He says that this Mosque is *in a little Village built to the Honour and under the Name of this Prophet*. Mr. *d' Herbelot* says noting of all this, tho' he relates many things reported of the *Mahometans* concerning *Jonas*. I will name a Modern Author, who, according to *Moreri*, made *a most ingenious Poem* upon the History of the Prophet.

From James Fenimore Cooper, *The Pilot; A Tale of the Sea* (New York: Charles Wiley, 1824) and *The Sea Lions; or, The Lost Sealers* (New York: Stringer & Townsend, 1849)

INTRODUCTION

James Fenimore Cooper (1789–1851) was America's first successful professional novelist and the creator of frontier hero Natty Bumppo in the five volumes of the Leatherstocking Tales. He invented the sea novel in response to what he thought was unjustified praise of Sir Walter Scott's (1771–1832) *The Pirate* (1819). In that work, Scott had based a scene in which Orkney islanders attacked a stranded whale on his own journal account. Scott, however, substituted a Right Whale for the dolphin-sized Blackfish he had himself seen. Cooper responded with a scene (itself rather implausible) of whaling from boats in *The Pilot* (1824). The whaling scene is one of three (including a naval combat and a shipwreck) that appear to deliberately illustrate the potential conflicts available to this new form of fiction. Melville claimed to have read Cooper "Long ago, and far inland," before reviewing Cooper on the eve of *Moby-Dick*. Bercaw # 167. Modern edition: State University of New York Press, 1986.

PRIMARY SOURCE

"By heaven, Tom," cried Barnstable, starting, "there is the blow of a whale!"

"Ay, ay, sir," returned the cockswain with undisturbed composure; "here is his spout not half a mile to seaward; the easterly gale has driven the creatur to leeward, and he begins to find himself in shoal water. He's been sleeping, while he should have been working to windward!"

"The fellow takes it coolly, too! he's in no hurry to get an offing!"

"I rather conclude, sir," said the cockswain, rolling over his tobacco in his mouth very composedly, while his little sunken eyes began to twinkle with pleasure at the sight, "the gentleman has lost his reckoning, and don't know which way to head to take himself back into blue water."

"Tis a finback!" exclaimed the lieutenant; "he will soon make headway, and be off."

"No, sir, 'tis a right-whale," answered Tom; "I saw his spout; he threw up a pair of as pretty rainbows as a Christian would wish to look at. He's a raal oil-butt, that fellow!"

Barnstable laughed, turned himself away from the tempting sight, and tried to look at the cliffs; and then unconsciously bent his longing eyes again on the sluggish animal, who was throwing his huge carcass, at times, for many feet from the water, in idle gambols. The temptation for sport, and the recollection of his early habits, at length prevailed over his anxiety in behalf of his friends, and the young officer inquired of his cockswain:

"Is there any whale-line in the boat, to make fast to that harpoon which you bear about with you in fair weather or foul?"

"I never trust the boat from the schooner without part of a shot, sir," returned the cockswain; "there is something nateral in the sight of a tub to my old eyes."

Barnstable looked at his watch, and again at the cliffs, when he exclaimed, in joyous tones:

"Give strong way, my hearties! There seems nothing better to be done; let us have a stroke of a harpoon at that impudent rascal."

The men shouted spontaneously, and the old cockswain suffered his solemn visage to relax into a small laugh, while the whale-boat sprang forward like a courser for the goal. During the few minutes they were pulling towards their game, long Tom arose from his crouching attitude in the stern-sheets, and transferred his huge form to the bows of the boat, where he made such preparations to strike the whale as the occasion required. The tub, containing about half of a whale-line, was placed at the feet of Barnstable, who had been preparing an oar to steer with in place of the rudder, which was unshipped, in order that, if necessary, the boat might be whirled round when not advancing.

Their approach was utterly unnoticed by the monster of the deep, who continued to amuse himself with throwing the water in two circular spouts high into the air, occasionally flourishing the broad flukes of his tail with a graceful but terrific force, until the hardy seamen were within a few hundred feet of him, when he suddenly cast his head downward and, without an apparent effort, reared his immense body for many feet above the water, waving his tail violently, and producing a whizzing noise, that sounded like the rushing of winds.

The cockswain stood erect, poising his harpoon, ready for the blow; but when he beheld the creature assume this formidable attitude, he waved his hand to his commander, who instantly signed to his men to cease rowing. In this situation the sportsmen rested a few moments, while the whale, struck several blows on the water in rapid succession, the noise of which re-echoed along the cliffs, like the hollow reports of so many cannon. After this wanton exhibition of his terrible strength, the monster sank again into his native element, and slowly disappeared from the eyes of his pursuers.

"Which way did he head, Tom?" cried Barnstable, the moment the whale was out of sight.

"Pretty much up and down, sir," returned the cockswain, whose eye was gradually brightened with the excitement of the sport; "he'll soon run his nose against the bottom if he stands long on that course, and will be glad to get another snuff of pure air; send her a few fathoms to starboard, sir, and I promise we shall not be out of his track."

The conjecture of the experienced old seaman proved true; for in a few moments the water broke near them, and another spout was cast into the air, when the huge animal rushed for half his length in the same direction, and fell on the sea with a turbulence and foam equal to that which is produced by the launching of a vessel, for the first time, into its proper element. After this evolution the whale rolled heavily, and seemed to rest for further efforts.

His slightest movements were closely watched by Barnstable and his cockswain, and when he was in a state of comparative rest, the former gave a signal to his crew to ply their oars once more. A few long and vigorous strokes sent the boat directly up to the broadside of the whale, with its bows pointing towards one of the fins, which was, at times, as the animal yielded sluggishly to the action of the waves, exposed to view. The cockswain poised his harpoon with much precision, and then darted it from him with a violence that buried the iron in the blubber of their foe. The instant the blow was made, long Tom shouted, with singular earnestness:

"Starn all!"

"Stern all!" echoed Barnstable; when the obedient seamen, by united efforts, forced the boat in a backward direction beyond the reach of any blow from their formidable antagonist. The alarmed animal, however, meditated no such resistance; ignorant of his own power, and of the insignificance of his enemies, he sought refuge in flight. One moment of stupid surprise succeeded the entrance of the iron, when he cast his huge tail into the air, with a violence that threw the sea around him into increased commotion, and then disappeared with the quickness of lightning, amid a cloud of foam.

"Snub him!" shouted Barnstable; "hold on, Tom; he rises already."

"Ay, ay, sir," replied the composed cockswain, seizing the line, which was running out of the boat with a velocity that rendered such a manoeuvre rather hazardous, and causing it to yield more gradually round the large loggerhead that was placed in the bows of the boat for that purpose. Presently the line stretched forward, and rising to the surface with tremulous vibrations, it indicated the direction in which the animal might be expected to reappear. Barnstable had cast the bows of the boat towards that point, before the terrified and wounded victim rose once more to the surface, whose time was, however, no longer wasted in his sports, but who cast the waters aside as he forced his way, with prodigious velocity, along the surface. The boat was dragged violently in his wake, and cut through the billows with a terrific rapidity, that at moments appeared to bury the slight fabric in the ocean. When long Tom beheld his victim throwing his spouts on high again, he pointed with exultation to the jetting fluid, which was streaked with the deep red of blood, and cried:

"Ay! I've touched the fellow's life! it must be more than two foot of blubber that stops my iron from reaching the life of any whale that ever sculled the ocean!"

"I believe you have saved yourself the trouble of using the bayonet you have rigged for a lance," said his commander, who entered into the sport with all the ardor of one whose youth had been chiefly passed in such pursuits: "feel your line, Master Coffin; can we haul alongside of our enemy? I like not the course he is steering, as he tows us from the schooner."

"'Tis the creatur's way, sir," said the cockswain; "you know they need the air in their nostrils, when they run, the same as a man; but lay hold, boys, and let's haul up to him."

The seamen now seized the whale-line, and slowly drew their boat to within a few feet of the tail of the fish, whose progress became sensibly less rapid, as he grew weak with the loss of blood. In a few minutes he stopped running, and appeared to roll uneasily on the water, as if suffering the agony of death.

"Shall we pull in, and finish him, Tom?" cried Barnstable; "a few sets from your bayonet would do it."

The cockswain stood examining his game with cool discretion, and replied to this interrogatory:

"No, sir, no—he's going into his flurry; there's no occasion for disgracing ourselves by using a soldier's weapon in taking a whale. Starn off, sir, starn off! the creater's in his flurry!"

The warning of the prudent cockswain was promptly obeyed, and the boat cautiously drew off to a distance, leaving to the animal a clear space, while under its dying agonies. From a state of perfect rest, the terrible monster threw its tail on high, as when in sport, but its blows were trebled in rapidity and violence, till all was hid from view by a pyramid of foam, that was deeply dyed with blood. The roarings of the fish were like the bellowing of a herd of bulls; and to one who was ignorant of the fact, it would have appeared as if a thousand monsters were engaged in deadly combat behind the bloody mist that obscured the view. Gradually, these effects subsided, and when the discolored water again settled down to the long and regular swell of the ocean, the fish was seen, exhausted, and yielding passively to its fate. As life departed, the enormous black mass rolled to one side; and when the white and glistening skin of the belly became apparent, the seamen well knew that their victory was achieved.

"What's to be done now?" said Barnstable, as he stood and gazed with a diminished excitement at their victim; "he will yield no food, and his carcass will probably drift to land, and furnish our enemies with the oil."

"If I had but that creatur in Boston Bay," said the cockswain, "it would prove the making of me; but such is my luck forever! Pull up, at any rate, and let me get my harpoon and line—the English shall never get them while old Tom Coffin can blow."

INTRODUCTION

Many years later, Cooper wrote this whaling scene for *The Sea Lions* to establish the respective characters of his protagonist Captain Roswell Gardiner and foil Captain Jason Daggett. In his religious allegory, Cooper has two identical schooners fitted out to pursue seals and buried treasure. Catalyzed by the efforts of evangelical seaman Stephen Stimson, Gardiner, through exposure to the natural world and the death of his alter ego Daggett, converts from Unitarianism to an acceptance of the Trinity. Melville reviewed *The Sea Lions* for the April 28 (1849) *Literary World*: "The reader is reminded of the appalling adventures of the United States Exploring Ship in the same part of the world as narrated by Wilkes, and of Scoresby's Greenland narrative," Melville wrote. Sealts #160; Bercaw #171.

PRIMARY SOURCE

Preface

. . . In this book the design has been to pourtray [*sic*] man on a novel field of action, and to exhibit his dependence on the hand that does not suffer a sparrow to fall unheeded. The recent attempts of science, which employed the seamen of the four greatest maritime states of Christendom, made discoveries that have rendered the polar circles much more familiar to this age, than to any that has preceded it, so far as existing records show. We say "existing records;" for there is much reason for believing that the ancients had a knowledge of our hemisphere, though less for supposing that they ever braved the dangers of the high latitudes. Many are, just at this moment, much disposed to believe that "Ophir" was on this continent; though for a reason no better than the circumstance of the recent discoveries of much gold. Such savans should remember that 'peacocks' came from ancient Ophir. If this be in truth that land, the adventurers of Israel caused it to be denuded of that bird of beautiful plumage.

Such names as those of Parry, Sabine, Ross, Franklin, Wilkes, Hudson, Ringgold, &c., &c., with those of divers gallant Frenchmen and Russians, command our most profound respect; for no battles or victories can redound more to the credit of seamen than the dangers they all encountered, and the conquests they have all achieved. One of those named, a resolute and experienced seaman, it is thought must, at this moment, be locked in the frosts of the arctic circle, after having passed half a life in the endeavour to push his discoveries into those remote and frozen regions. He bears the name of the most distinguished of the philosophers of this country; and nature has stamped on his features—by one of those secret laws which just as much baffle our means of comprehension, as the greatest of all our mysteries, the incarnation of the Son of God—a resemblance that, of itself, would go to show that they are of the same race. Any one who has ever seen this emprisoned [*sic*] navigator, and who is familiar with the countenances of the men of the same name who are to be found in numbers amongst ourselves, must be struck with a likeness that lies as much beyond the grasp of that reason of which we are so proud, as the sublimest facts taught by induction, science, or revelation. Parties are, at this moment, out in search of him and his followers; and it is to be hoped that the Providence which has so singularly attempered the different circles and zones of our globe, placing this under a burning sun, and that beneath enduring frosts, will have included in its divine forethought a sufficient care for these bold wanderers to restore them, unharmed, to their friends and country. In a contrary event, their names must be transmitted to posterity as the victims to a laudable desire to enlarge the circle of human knowledge, and with it, we trust, to increase the glory due to God. . . .

Chapter X

. . . Although neither of the schooners was thoroughly fitted for a whaler, each had lines, lances, harpoons, &c., in readiness in their quarter-boats, prepared for any turn of luck like this which now offered. The process of paddling up to whales, which is now so common in the American ships, was then very little or not at all resorted to. It is said that the animals have got to be so shy, in consequence of being so much pursued, that the old mode of approaching them will not suffice, and that it now requires much more care and far more art to take one of these creatures, than it did thirty years since. On this part of the subject, we merely repeat what we hear, though we think we can see an advantage in the use of the paddle that is altogether independent of that of the greater quiet of that mode of forcing a boat ahead. He that paddles looks *ahead,* and the approach is more easily regulated, when the whole of the boat's crew are apprised, by means of their own senses, of the actual state of things, than when they attain their ideas of them through the orders of an officer. The last must govern in all cases, but the men are prepared for them, when they can see what is going on, and will be more likely to act with promptitude and intelligence, and will be less liable to make mistakes.

The four boats, two from each schooner, dropped into the water nearly about the same time. Daggett was at the steering-oar of one, as was Roswell at that of another. Hazard, and Macy, the chief mate of the Vineyard craft, were at the steering-oars of the two remaining boats. All pulled in the direction of the spot on the ocean where the spouts had been seen. It was the opinion of those who had been aloft, that there were several *fish*; and it was certain that they were of the most valuable species, or the spermaceti, one barrel of the oil of which was worth about as much as the oil of three of the ordinary sort, or that of the *right* whale, supposing them all to yield the same quantity in number of barrels. The nature or species of the fish was easily enough determined by the spouts; the right whale throwing up two high arched jets of water, while the spermaceti throws but a single, low, bushy one.

It was not long ere [*sic*] the boats of the two captains came abreast of each other, and within speaking distance. A stern rivalry was now apparent in every countenance, the men pulling might and main, and without even a smile among them all. Every face was grave, earnest, and determined; every arm strung to its utmost powers of exertion. The men rowed beautifully, being accustomed to the use of their long oars in rough water, and in ten minutes they were all fully a mile dead to windward of the two schooners.

Few things give a more exalted idea of the courage and ingenuity of the human race than to see adventurers set forth, in a mere shell, on the troubled waters of the open ocean, to contend with and capture an animal of the size of the whale. The simple circumstance that the last is in its own element, while its assailants are compelled to approach it in such light and fragile conveyances, that, to the unpractised eye, it is sufficiently difficult to manage them amid the rolling waters, without seeking so powerful an enemy to contend with. But, little of all this did the crews of our four boats now think. They had before them the objects, or *one* of the objects, rather, of their adventure, and so long as that was the case, no other view but that of prevailing could rise before their eyes.

"How is it, Gar'ner?" called out the Vineyard master; "shall it be shares? or does each schooner whale on her own hook?"

This was asked in a friendly way, and apparently with great indifference as to the nature of the reply, but with profound art. It was Daggett's wish to establish a sort of partnership, which, taken in connection with the good feeling created by the affair at Beaufort, would be very apt to lead on to further and more important association. Luckily for Gardiner, an idea crossed his mind, just as he was about to reply, which induced the wisest answer. It was the thought, that competition would be more likely to cause exertion than a partnership, and that the success of all would better repay them for their toils and risks, should each vessel act exclusively for itself. This is the principle that renders the present state of society more healthful and advantageous than that which the friends of the different systems of associating, that are now so much in vogue, wish to substitute in its place. Individuality is an all-important feeling in the organization of human beings into communities; and the political economist who does not use it as his most powerful auxiliary in advancing civilization, will soon see it turn round in its tracks, and become a dead weight; indulging its self-love, by living with the minimum of exertion, instead of pushing his private advantage, with the maximum.

"I think each vessel had better work for herself and her owners," answered Roswell Gardiner.

As the schooners were in the trades, there was a regular sea running, and one that was neither very high nor much broken. Still, the boats were lifted on it like egg-shells or bubbles, the immense power of the ocean raising the largest ships, groaning under their vast weight of ordnance, as if they were feathers. In a few minutes, Gardiner and Daggett became a little more separated, each looking eagerly for the spouts, which had not been seen by either since quitting his vessel. All this time the two mates came steadily on, until the whole of the little fleet of boats was, by this time, not less than a marine league distant from the schooners. The vessels themselves were working up to windward, to keep as near to the boats as possible, making short tacks under reduced canvass; a shipkeeper, the cook, steward, and one or two other hands, being all who were left on board them.

We shall suppose that most of our readers are sufficiently acquainted with the general character of that class of animals to which the whale belongs, to know that all of the genus breathe the atmospheric air, which is as necessary for life to them as it is to man himself. The only difference in this respect is, that the whale can go longer without renewing his respiration than all purely land-animals, though he must come up to breathe at intervals, or die. It is the exhaling of the old stock of air, when he brings the "blow-holes," as seamen call the outlets of his respiratory organs, to the surface, that forces the water upward, and forms the "spouts," which usually indicate to the whalers the position of their game. The "spouts" vary in appearance, as has been mentioned, owing to the number and situation of the orifices by which the exhausted air escapes. No sooner is the vitiated air exhaled, than the lungs receive a new supply; and the animal either remains near the surface, rolling about and sporting amid the waves, or descends again, a short distance, in quest of its food. This food, also, varies materially in the different species. The right-whale is supposed to live on what may be termed marine insects, or the molluscæ of the ocean, which it is thought he obtains by running in the parts of the sea where they most abound; arresting them by the hairy fibres which grow on the laminæ of bone that, in a measure, compose his jaws, having no teeth. The spermaceti, however, is furnished with regular grinders, which he knows very well how to use, and with which he often crushes the boats of those who come against him. Thus, the whalers have but one danger to guard against, in assaulting the common animal, viz., his flukes, or tail; while the spermaceti, in addition to the last means of defence, possesses those of his teeth or jaws. As this latter animal is quite one-third head, he has no very great dissemblance to the alligator in this particular.

By means of this brief description of the physical formation and habits of the animals of which our adventurers were in pursuit, the general reader will be the better able to understand that which it is our duty now to record. After rowing the distance named, the boats became a little separated, in their search for the fish. That spouts had been seen, there was no doubt; though, since quitting the schooners, no one in the boats had got a further view of the fish,—if fish, animals with respiratory organs can be termed. A good look-out for spouts had been kept by each man at the steering-oars, but entirely without success. Had not Roswell and Daggett, previously to leaving their respective vessels, seen the signs of whales with their own eyes, it is probable that they would now have both been disposed to return, calling in their mates. But, being certain that the creatures they sought were not far distant, they continued slowly to separate, each straining his eyes in quest of his game, as his boat rose on the summit of the rolling and tossing waves. Water in motion was all around them; and the schooners working slowly up against the trades, were all that rewarded their vigilant and anxious looks. Twenty times did each fancy that he saw the dark back, or head, of the object he sought; but as often did it prove to be no more than a lipper of water, rolling up into a hummock ere it broke, or melted away again into the general mass of the unquiet ocean. When it is remembered that the surface of the sea is tossed into a thousand fantastic outlines, as its waves roll along, it can readily be imagined how such mistakes could arise.

At length Gardiner discerned that which his practised eye well knew. It was the flukes, or extremity of the tail of an enormous whale, distant from him less than a quarter of a mile, and in such a position as to place the animal at about the same breadth of water from Daggett. It would seem that both of these vigilant officers perceived their enemy at the same instant, for each boat started for it, as if it had been instinct with life. The pike or the shark could not have darted towards its prey with greater promptitude, and scarcely with greater velocity, than these two boats. Very soon the whole herd was seen, swimming along against the wind, an enormous bull-whale leading, while half a dozen calves kept close to the sides of their dams, or sported among themselves, much as the offspring of land animals delight in their youth and strength. Presently a mother rolled lazily over on her side, permitting its calf to suck. Others followed this example; and then the leader of the herd ceased his passage to windward, but began to circle the spot, as if in complaisance to those considerate nurses who thus waited on the wants of their young. At this interesting moment, the boats came glancing in among the herd.

Had the competition and spirit of rivalry been at a lower point among our adventurers than it actually was, greater caution might have been observed. It is just as dangerous to assault a whale that has its young to defend, as to assault most other animals. We know that the most delicate women become heroines in such straits; and nature seems to have given to the whole sex, whether endowed with reason or only with an instinct, the same disposition to die in defence of the helpless creatures that so much depend on their care. But, no one there now thought of the risk he ran, it being the Vineyard against Oyster Pond, one Sea Lion against the other, and, in many instances, pocket against pocket.

Roswell, as if disdaining all meaner game, pulled quite through the herd, and laid the bows of his boat directly on the side of the old bull—a hundred-barrel whale, at the very least. No sooner did the enormous creature feel the harpoon, than, throwing its flukes upward, it descended into the depths of the ocean, with a velocity that caused smoke to arise from the chuck through which the line passed. Ordinarily, the movement of a whale is not much faster than an active man can walk; and, when it runs on the surface, its speed seldom exceeds that of a swift vessel under full sail; but, when suddenly startled, with the harpoon in its blubber, the animal is capable of making a prodigious exertion. When struck, it usually 'sounds,' as it is termed, or runs downward, sometimes to the depth of a mile; and it is said that instances have been known in which the fish inflicted great injury on itself, by dashing its head against rocks.

In the case before us, after running out three or four hundred fathoms of line, the 'bull' to which Gardiner had 'fastened,' came up to the surface, 'blowed,' and began to move slowly towards the herd again. No sooner was the harpoon thrown, than a change took place in the disposition of the crew of the boat, which it may be well to explain. The harpoon is a barbed javelin, fastened to a staff to give it momentum. The line is attached to this weapon, the proper use of which is to 'fasten' to the fish, though it sometimes happens that the animal is killed at the first blow. This is when the harpoon has been hurled by a very skilful and vigorous harpooner. Usually, this weapon penetrates some distance into the blubber in which a whale is encased, and when it is drawn back by the plunge of the fish, the barbed parts get embedded in the tough integuments of the hide, together with the blubber, and hold. The iron of the harpoon being very soft, the shank bends under the strain of the line, leaving the staff close to the animal's body. Owing to this arrangement, the harpoon offers less resistance to the water, as the whale passes swiftly through it. No sooner did the boat-steerer, or harpooner, cast his 'irons,' as whalers term the harpoon, than he changed places with Roswell, who left the steering-oar, and proceeded forward to wield the lance, the weapon with which the victory is finally consummated. The men now 'peaked' their oars, as it is termed; or they placed the handles in cleets [*sic*] made to receive them, leaving the blades elevated in the air, so as to be quite clear of the water. This was done to get rid of the oars, in readiness for other duty, while the instruments were left in the tholes, to be resorted to in emergencies. This gives a whale-boat a peculiar appearance, with its five long oars raised in the air, at angles approaching forty-five degrees. In the mean time, as the bull approached the herd, or school,[1] as the whalers term it, the boats' crew began to haul in line, the boat-steerer coiling it away carefully, in a tub placed in the stern-sheets purposely to receive it. Any one can understand how important it was that this part of the duty should be well performed, since bights of line running out of a boat, dragged by a whale, would prove so many snares to the men's legs, unless previously disposed of in a place proper to let it escape without this risk. For this reason it is, that the end of a line is never permitted to run out at the bow of a boat at all. It might do some injury in its passage, and an axe is always applied near the bows, when it is found necessary to cut from a whale.

It was so unusual a thing to see a fish turn towards the spot where he was struck, that Roswell did not know what to make of this manoeuvre in his bull. At first he supposed the animal meant to make fight, and set upon him with its tremendous jaws; but it seemed that caprice or alarm directed the movement; for, after coming within a hundred yards of the boat, the creature turned and commenced sculling away to windward, with wide and nervous sweeps of its formidable flukes. It is by this process that all the fish of this genus force their way through the water, their tails being admirably adapted to the purpose. As the men had showed the utmost activity in hauling in upon the line, by the time the whale went off to windward again they had got the boat up within about four hundred feet of him.

Now commenced a tow, dead to windward, it being known that a fish, when struck, seldom runs at first in any other direction. The rate at which the whale moved was not at the height of his speed, though it exceeded six knots. Occasionally, this rate was lessened, and in several instances his speed was reduced to less than half of that just mentioned. Whenever one of these lulls occurred, the men would haul upon the line, gradually getting nearer and nearer to the fish, until they were within fifty feet of his tremendous flukes. Here, a turn was taken with the line, and an opportunity to use the lance was waited for.

Whalers say that a forty-barrel bull of the spermaceti sort is much the most dangerous to deal with of all the animals of this species. The larger bulls are infinitely the most powerful, and drive these half-grown creatures away in herds by themselves, that are called 'pads,' a circumstance that probably renders the young bull discontented and fierce. The last is not only more active than the larger animal, but is much more disposed to make fight, commonly giving his captors the greatest trouble. This may be one of the reasons why Roswell Gardiner now found himself towing at a reasonable rate, so close upon the flukes of a hundred-barrel whale. Still, there was that in the movements of this animal, that induced our hero to be exceedingly wary. He was now two leagues from the schooners, and half that distance from the other boats, neither of which had as yet fastened to a fish. This latter circumstance was imputed to the difficulty the different officers had in making their selections, cows, of the spermaceti breed, when they give suck, being commonly light, and yielding, comparatively, very small quantities of head-matter and oil. In selecting the bull, Roswell had shown his judgment, the male animal commonly returning to its conquerors twice the profit that is derived from the female.

The whale to which Roswell was fast continued sculling away to windward for quite two hours, causing the men to entirely lose sight of the other boats, and bringing the topsails of the schooners themselves down to the water's edge. Fortunately, it was not yet noon, and there were no immediate apprehensions from the darkness; nor did the bull appear to be much alarmed, though the boat was towing so close in his rear. At first, or before the irons were thrown, the utmost care had been taken not to make a noise; but the instant the crew were 'fast,' whispers were changed into loud calls, and orders were passed in shouts, rather than in verbal commands. The wildest excitement prevailed among the men, strangely blended with a cool dexterity; but it was very apparent that a high sporting fever was raging among them. Gardiner himself was much the coolest man in his own boat, as became his station and very responsible duties.

Stimson, the oldest and the best seaman in the schooner, he who had admonished his young commander on the subject of the gratitude due to the Deity, acted as the master's boat-steerer, having first performed the duty of harpooner. It was to him that Gardiner now addressed the remarks he made, after having been fastened to his whale fully two hours.

"This fellow is likely to give us a long drag," said the master, as he stood balancing himself on the clumsy cleets in the bows of the boat, using his lance as an adept in saltation poises his pole on the wire, the water curling fairly above the gunwale forward, with the rapid movement of the boat; "I would haul up alongside, and give him the lance, did I not distrust them flukes. I believe he knows we are here."

"That he does—that does he, Captain Gar'ner. It's always best to be moderate and wait your time, sir. There's a jerk about that chap's flukes that I don't like myself, and it's best to see what he would be at, before we haul up any nearer. Don't you see, sir, that every minute or two he strikes down, instead of sculling off handsomely and with a wide sweep, as becomes a whale?"

"That is just the motion I distrust, Stephen, and I shall wait a bit to see what he would be at. I hope those ship-keepers will be busy, and work the schooners well up to windward before it gets to be dark. Our man is asleep half his time, and is apt to let the vessel fall off a point or two."

"Mr. Hazard gave him caution to keep a bright look out, sir, and I think he'll be apt to—look out, sir!—Look out!"

This warning was well-timed; for, just at that instant the whale ceased sculling, and lifting its enormous tail high in the air, it struck five or six blows on the surface of the water, that made a noise which might have been heard half a league, besides filling the atmosphere immediately around him with spray. As the tail first appeared in the air, line was permitted to run out of the boat, increasing the distance between its bows and the flukes to quite a hundred feet. Nothing could better show the hardy characters of the whalers than the picture then presented by Roswell Gardiner and his companions. In the midst of the Atlantic, leagues from their vessel, and no other boat in sight, there they sat patiently waiting the moment when the giant of the deep should abate in his speed, or in his antics, to enable them to approach and complete their capture. Most of the men sat with their arms crossed, and bodies half-turned, regarding the scene, while the two officers, the master and boat-steerer, if the latter could properly be thus designated, watched each evolution with a keenness of vigilance that let nothing like a sign or a symptom escape them.

Such was the state of things, the whale still threshing the sea with his flukes, when a cry among his men induced Roswell for a moment to look aside. There came Daggett fast to a small bull, which was running directly in the wind's eye with great speed, dragging the boat after him, which was towing astern at a distance of something like two hundred fathoms. At first, Roswell thought he should be compelled to cut from his whale, so directly towards his own boat did the other animal direct his course. But, intimidated, most probably, by the tremendous blows with which the larger bull continued to belabour the ocean, the smaller animal sheered away in time to avoid a collision, though he now began to circle the spot where his dreaded monarch lay. This change of course gave rise to a new source of apprehension. If the smaller bull should continue to encircle the larger, there was great reason to believe that the line of Daggett might get entangled with the boat of Gardiner, and produce a collision that might prove fatal to all there. In order to be ready to meet this danger, Roswell ordered his crew to be on the look-out, and to have their knives in a state for immediate use. It was not known what might have been the consequence of this circular movement as respects the two boats; for, before they could come together, Daggett's line actually passed into the mouth of Gardiner's whale, and drawing up tight into the angle of his jaws, set the monster in motion with a momentum and power that caused the iron to draw from the smaller whale, which by this time had more than half encircled the animal. So rapid was the rate of running now, that Roswell was obliged to let out line, his whale sounding to a prodigious depth. Daggett did the same unwilling to cut as long as he could hold on to his line.

At the expiration of five minutes the large bull came up again for breath, with both lines still fast to him; the one in the regular way, or attached to the harpoon, and the other jammed in the jaws of the animal by means of the harpoon and staff, which formed a sort of toggle at the angle of his enormous mouth. In consequence of feeling this unusual tenant, the fish compressed its jaws together, thus rendering the fastening so much the more secure. As both boats had let run line freely while the whale was sounding, they now found themselves near a quarter of a mile astern of him, towing along, side by side, and not fifty feet asunder. If the spirit of rivalry had been aroused among the crew of these two boats before, it was now excited to a degree that menaced acts of hostility.

"You know, of course, Captain Daggett, that this is my whale," said Gardiner. "I was fast to him regularly, and was only waiting for him to become a little quiet to lance him, when your whale crossed his course, fouled your line, and has got you fast in an unaccountable way, but not according to whaling law."

"I don't know that. I fastened to a whale, Captain Gar'ner, and am fast to a whale now. It must be *proved* that I have no right to the crittur' before I give him up."

Gardiner understood the sort of man with whom he had to deal too well to waste words in idle remonstrances. Resolved to maintain his just rights at every hazard, he ordered his men to haul in upon the line, the movement of the whale becoming so slow as to admit of this measure. Daggett's crew did the same, and a warm contest existed between the two boats, as to which should now first close with the fish and kill it. This was not a moment for prudence and caution. It was "haul in—haul in, boys," in both boats, without any regard to the danger of approaching the whale. A very few minutes sufficed to bring the parties quite in a line with the flukes, Gardiner's boat coming up on the larboard or left-hand side of the animal, where its iron was fast, and Daggett's on the opposite, its line leading out of the jaws of the fish in that direction. The two masters stood erect on their respective clumsy cleets [*sic*], each poising his lance, waiting only to get near enough to strike. The men were now at the oars, and without pausing for any thing, both crews sprung to their ashen instruments, and drove the boats headlong upon the fish. Daggett, perhaps, was the coolest and most calculating at that moment, but Roswell was the most nervous, and the boldest. The boat of the last actually hit the side of the whale, as its young commander drove his lance through the blubber, into the vitals of the fish. At the same instant Daggett threw his lance with consummate skill, and went to the quick. It was now "stern all!" for life, each boat backing off from the danger as fast as hands could urge. The sea was in a foam, the fish going into his "flurry" almost as soon as struck, and both crews were delighted to see the red of the blood mingling its deep hues with the white of the troubled water. Once or twice the animal spouted, but it was a fluid dyed in his gore. In ten minutes it turned up and was dead.

Note

1. We suppose this word to be a corruption of the dutch "*schule*" which, we take it, means the same thing.

From Nathaniel Ames, *A Mariner's Sketches*

(Providence: Cory, Marshall, and Hammond, 1830)

INTRODUCTION

Not much is known of Ames (1796–1835), but Melville is likely to have seen Ames's book as early as 1849, since traces of it show up in *Redburn* and *White-Jacket*. In each of three books, starting with this one, Ames expresses his extreme dislike of the writings of James Fenimore Cooper—a view not shared by Melville in his reviews of *The Sea Lions* (1849) and *The Red Rover* (1850). See Charles Anderson, "Cooper's Sea Novels Spurned in the Maintop," *Modern Language Notes* 66:6 (June 1951), 388–391, and Warren Walker, "Ames vs. Cooper: the Case Re-Opened," *Modern Language Notes* 70:1 (January 1955), 27–32. Bercaw #13.

PRIMARY SOURCE

There was nothing more striking about the New South Shetland Islands, than the prodigious number of whales. They were principally of the two species known by the names of 'hump-backs' and 'right' whales, though 'fin-backs' were occasionally seen, and sperm whales teeth were found on many of the beaches. Besides whales there were a few black fish, and among the smaller fry, were a very few 'snappers' a species of fish of the size of pearch, and in the stomachs of the seals are sometimes found shrimps.

Of birds there were plenty, gulls, albatrosses, twelve or fourteen feet from the tip of one wing to the tip of the other, Cape pigeons and mother Carey's chickens. It is confidently asserted, and firmly believed by the older 'salts,' that these chickens lay their eggs at sea, and hatch them under their wings, and when we found the cliffs filled with their nests and young, those who held the established and orthodox belief, insisted that the South Shetland mother careys were a distinct race from those on the Grand Banks, and in the Gulf Stream.

Besides we had a kind of bird like a pigeon, in size and taste, but web-footed; and last but by no means least, penguins in 'numbers without number.' These birds are somewhat larger than a goose, but in place of wings have two short flippers, and stand upright on their legs, which are placed further back than in other birds, so that at a little distance they resemble young children with white aprons. Notwithstanding this apparently awkward conformation they would make shift to get up rocks where we could not follow them. They make their nests of small pebbles, in immensely numerous companies, nearly adjoining one another, and are continually stealing stones from each others [*sic*] nests, and as continually fighting. . . .

Just before a gale of wind their cackling is deafening, and our men used to say when they were peculiarly noisy, 'ah d—n you! you are praying for another gale of wind, are you?' The majority of our men firmly believed that they had formerly been human beings, who for 'deeds done in the body' were condemned to do penance in the shape of a goose, in this cold and barren region. The males they set down as quondam sea captains and tyrannical navy officers, and the females as Liverpool and Wapping landladies &c.

I have observed generally, that sailors are rather inclined to the Pythagorean philosophy, and hold to the transmigration of souls. One of their proverbs, namely, 'every dog must have his day, and every b—h two Sunday afternoons,' is commonly explained by supposing that in another world; captains and mates will go to sea before the mast and so *vice versa*. An hypothesis as rational, and a thousand times more humane than supposing that there are 'infants in hell not a span long,' or that one of the chief joys of paradise is to 'look down from heaven and behold the unutterable and everlasting torments of the wicked,' which no doubt would be extremely pleasant to a vast many zealous saints of my acquaintance, who would be delighted if they had the power to employ the secular arm to give sinners a foretaste, in this world, of those pains which are to constitute, in the next, one of the most entertaining spectacles to the 'spirits of just men made perfect.'

Elephant blubber, as above mentioned, constituted almost our sole fuel, but sometimes we found the bones of whales full of oil, which burned with great vehemence. There was scarcely a beach that had not a greater or less portion of a whale's skeleton on it, and in one place we found the entire frame of a very large whale, nearly a half a mile above high water mark. We also found and measured a jaw bone, that was twenty-two feet and six inches in length. An old Nantucket whaling captain who saw it, declared it to be by far the largest he had ever seen.

As for provisions, our bill of fare consisted of the liver, heart and flippers or fins of the seal, which parboiled and fried were excellent, 'elephants' tongues, far superior to neats', and several different species of birds. Eggs we had in plenty in the beginning of the season, so that we lived at least comfortably, and were always blessed with huge appetites. In short we had become so attached to this roving, free and unconfined life, that a dozen or fourteen of us volunteered to stay all winter, a motion that was 'over-ruled' by the captain.

From "Whales," *Penny Cyclopædia of the Society for the Diffusion of Useful Knowledge*

Vol. 27 (1843)

INTRODUCTION

Charles Knight (1791–1873) was a literary entrepreneur of unbounded energy and, as it turns out, great redundancy. The same or slightly altered text might appear in Knight's *Penny Magazine* (1832–1845), the *Penny Cyclopædia* (1833–1843), or a separate volume also published by Knight. He is perhaps best known today as the editor of *The Pictorial Shakespeare* (1842), which, like many of his other publications, was initially published in parts to make it affordable to the working classes.

Kendra Gaines initiated the study of Melville's use of Knight's 30-volume encyclopedia, whose entries were often thinly adapted passages from authoritative works. Subsequent scholars have gone on to identify traces of the *Penny Cyclopædia* in other works, particularly *Israel Potter.* Beale, Kitto, and Scoresby were among the contributors, and it can be difficult—or impossible—to tell whether Melville is using the *Penny Cyclopædia* or the original source for any specific borrowing. In this selection, the reader can see how transparent Knight's borrowings were. Knight's illustrations, also derivative and mostly of skeletal remains, are not reproduced here. Melville is likely to have consulted the *Penny Cyclopædia* as early as spring 1849.

Although obviously important as a starting point for Melville's investigations of whaling per se, the *Penny Cyclopædia* is likely to have provided Melville with the link to George L. Craik's *The New Zealanders* (1830), the work that gave Melville the material he used to develop the middle stage of *Moby-Dick*'s outlandish harpooner Queequeg as a New Zealander—after being a square-tattooed Native American but before becoming a native of the imaginary Kokovoko. See Lucy Maddox, *Removals: Nineteenth-Century American Literature and the Politics of Indian Affairs* (New York: Oxford University Press, 1991), p. 64 and Geoffrey Sanborn, "Whence Come You Queequeg?" *American Literature* 77:2 (June 2005), 227–257. Bercaw #544.]

PRIMARY SOURCE

The cetaceous mammals, whose abode is either in the sea or the great rivers, resemble the FISHES so closely in external appearance, that it is hardly to be wondered at that not only the vulgar, but even some of the earlier zoologists looked upon them as belonging to that class. This notion is kept alive to the present day in the announcements of the comparative success of those ships which are employed in the *Whale Fishery*; for not only is it conveyed by that general term for the capture of whales, but by statements that one ship has arrived with three *fish*, another with four *fish*, a third with one *fish*, &c. . . .

The following is the Linnæan definition of the last-named order:—

Pectoral fins in lieu of feet, and feet conjoined into a horizontal flattened fin in lieu of a tail. No claws. Teeth cartilaginous. Nose often a frontal pipe. Food, mollusks, fishes.

Locality, the Ocean.

Linnæus then declares that he has separated these cetaceans from the fishes, and associated them with the mammals, on account of their warm bilocular heart, their lungs, their moveable eyelids, their hollow ears, *penem intrantem feminam mammis lactantem*, and this, to use his own expressive words, *ex lege naturae jure meritoque.*

Here then we find the decisive step taken, with the unflinching firmness of a master mind, relying upon the philosophical principles that demanded the separation, and no longer yielding to popular prejudice by calling that a fish which he knew to be a mammiferous animal. Some parts of his definition—not much of it—may be open to criticism, as where he designates the teeth as cartilaginous, a term probably used to comprehend both the horny laminæ of the whalebone whales and the true teeth of the other cetaceans; but the broad line of distinction is unassailable, and will ever remain so. . . .

Physeter

Generic Character.—Length of the head equal to a third or half of the total length; upper jaw large, elevated, and either without teeth or furnished with very few, which are short, and nearly entirely hidden by the gum; lower jaw narrow, and armed with stout conical teeth; orifices of the blow-holes united, and so situated at the end of the upper part of the muzzle; no dorsal fin.

The subjoined cut is from M. F. Cuvier, who gives it from the skeleton in the Paris museum, and is confined to the lower jaw only, from which it may be inferred that in the French specimen there is no appearance of teeth in the upper jaw: in the lower there are 27 on each side=54.

Example *Physeter macrocephalus.*

Description.—To render the following abridgment of the description by Mr. Beale (who, in his excellent work on the 'Natural History of the Sperm Whale,'[1] has done more to elucidate its habits and form than any other writer), more intelligible, we prefix, as he himself does, his cut, which is by far the most accurate published figure extant of the Spermaceti Whale.

The head presents a very thick blunt extremity, constituting about a third of the whole length of the animal; at its junction with the body is a large protuberance on the back called 'the bunch of the neck;' immediately behind this is the thickest part of the body, which from thence gradually tapers off to the tail, but it does not become much smaller for about another third of the whole length, when 'the small' or tail commences; and at this point also, on the back, is a large pyramidal prominence called 'the hump,' from which a series of smaller processes run half way down the 'small' or tail, constituting what the whalers term the 'ridge.' The body then contracts so much as to become finally not thicker than that of a man, and terminates by expanding on the sides into the 'flukes' or tail, forming a large triangular horizontal fin with a slight notch or depression posteriorly between the flukes, which are about six or eight feet in length, and from twelve to fourteen in breadth in the largest males or 'Bulls.' The chest and belly are narrower than the broadest part of the back, and taper off evenly towards the tail: the depth of the head and body is, in all parts except the tail, greater than the width. The head, viewed in front, presents a broad somewhat flattened surface, rounded and contracted above, considerably expanded on the sides, and gradually contracted below, resembling in some degree the cutwater of a ship. The slit of the single blowing-hole or nostril is about twelve inches in length. In the right side of the nose is the 'case,' a cavity for the purpose of secreting and containing an oily fluid, which after death concretes into a granulated yellowish substance: this is the spermaceti. In the case of a large whale there is not unfrequently a ton, or more than ten barrels, of spermaceti. Beneath the case and nostril is the elastic 'junk,' formed of dense cellular tissue strengthened by strong tendinous fibres, and infiltrated with very fine sperm oil and spermaceti. The mouth extends nearly the whole length of the head. Both the jaws, especially the lower, are contracted in front to a very narrow point; and when the mouth is closed the lower jaw is received within a sort of cartilaginous lip, or projection of the upper one; but principally in front, for further back at the sides, and towards the angle of the mouth, both jaws are furnished with tolerably well-developed lips. The tongue is small and white. The throat is capacious enough to give passage to the body of a man, presenting a strong contrast to the contracted gullet of the Greenland Whale. Throughout, the mouth is lined with a pearly white membrane. The eyes are small in proportion to the size of the animal, and are furnished with eyelids, the lower of which is most moveable. At a short distance behind the eyes are the external openings of the ears, sufficiently large to admit a small quill. Not far from the posterior angle of the mouth are the swimming-paws or fins, which are not much used in progression, but probably more as balances, and occasionally in supporting the young.

Mr. Beale gives the following as the dimensions of a sperm whale of the largest size, or about eighty-four feet in length:—depth of head from eight to nine feet; breadth from five to six feet; depth of body seldom exceeding twelve or fourteen feet; circumference seldom exceeding thirty-six feet; swimming-paws about six feet long and three broad.

The skin is smooth, but occasionally in old whales wrinkled. The general colour is very dark, deepest on the upper part of the head, back, and flukes, in which situation it is sometimes black; on the sides it gradually assumes a lighter tint, and on the breast becomes silvery grey. In different individuals there is however every variety of shade, and some are piebald. Old 'Bulls' have generally a portion of grey on the nose immediately above the fore-part of the upper jaw, when they are said to be grey-headed. The 'black skin' in young whales is about three-eighths of an inch thick: in old ones it is not more than one eighth. Immediately beneath the black skin is the blubber or fat, termed the 'blanket,' of a light yellowish colour, producing when melted the sperm oil.

Habits, Food, &c.—The bulk of the head is, as we have seen, made up of a membranous 'case,' containing a thin oil of much less specific gravity than water; below which again is the 'junk,' which, although heavier than the spermaceti, is still lighter than the element in which the whale moves; consequently, observes Mr. Beale, the head taken as a whole is lighter specifically than any other part of the body, and will always have a tendency to rise at least so far above the surface as to elevate the nostril or blow-hole sufficiently for all purposes of respiration; and more than this, a very slight effort on the part of the whale would only be necessary to raise the whole of the anterior flat surface of the nose out of the water. At very regular intervals of time the snout emerges, and from the extremity of the nose the spout is thrown up, and at a distance appears think, low, bushy, and white; it is formed of the expired air forcibly ejected through the blow-hole, and acquires its white colour from minute particles of water previously lodged in the chink or fissure of the nostril, and also from the condensation of the aqueous vapour thrown off by the lungs. The spout, says Mr. Beale in continuation, is projected at an angle of 136° in a slow and continuous manner for about three minutes, and may be seen from the mast-head in favourable weather at the distance of four or five miles. When the whale is alarmed or 'gallied,' the spout is thrown much higher with great rapidity, and differs much from its usual appearance. Immediately after each spout the nose sinks beneath the water, scarcely a second intervening for the act of inspiration, which must consequently be performed very quickly, the air rushing into the chest with astonishing velocity; there is however no sound caused by inspiration, and very little by expiration in this species; in short, nothing of that loud noise called the 'drawback' in the Finback and other whales. Ten seconds is occupied by a large bull sperm whale in making one inspiration and one expiration: during six of these the nostril is beneath the water. At each breathing time the whale makes from sixty to seventy expirations, and remains therefore at the surface ten or eleven minutes. When the breathing-time is over, or, as the whalers term it, he has had his 'spoutings out,' the head sinks slowly, the 'small,' or the part between the 'hump' and 'flukes' appears above the water curved, with the convexity upwards; the flukes are then lifted high into the air, and the animal having assumed a straight position, descends perpendicularly to an unknown depth: this last act is called 'peaking the flukes,' and those who are on the lookout call loudly when they see it—'there goes flukes. The whale continues thus hidden beneath the surface for one hour and ten minutes; some will remain one hour and twenty minutes, and others only for one hour; but these, Mr. Beale says, are rare exceptions. A seventh of the time of this whale is, Mr. Beale makes out, consumed in respiration.

Small fishes are occasionally swallowed in quantities by this whale, and one has been known to eject from its stomach a fish as large as a moderate sized salmon; but the principal food of the Sperm Whale appears to consist of squids or cuttle-fishes. [SEPIADÆ.]

This species is gregarious; and the herds called 'schools' are of two kinds, one consisting of females, the other of young males not fully grown. Mr. Beale has seen as many as five or six hundred in one 'school.' With each female 'school' are from one to three large 'bulls' or 'schoolmasters,' as they are termed by the whalers. The full-grown males almost always go alone in search of food: they are when alone very incautious and easily killed. It is the smaller, or 'forty-barrel bull,' as he is called, that makes the most desperate resistance. A large whale will yield eighty, and sometimes one hundred barrels of oil.

Mr. Beale states that the female is smaller than the male, and that she breeds at all seasons, producing generally only one at a time, but sometimes two. Nothing certain appears to be known as to the period of gestation, but M. F. Cuvier supposes it to be ten months. A fœtal cachalot, dissected by Mr. Bennett,[2] was fourteen feet long and six in circumference, deep black mottled with white spots. Its position in the womb was that of a bent bow. According to M. F. Cuvier, the two brought forth by the stranded whale near D'Audierne, were ten or eleven feet long; and Captain Colnett states that the young sperm whales he saw in great numbers off the Galapagos Islands were not larger than a small porpoise. Mr. Beale's own observations coincided with those of Mr. Bennett.

For many other habits of this whale, such as 'breaching,' or leaping clear out of the water and falling back again on its side, so that the breach may be seen in a clear day from the mast-head at a distance of six miles; 'going head out,' a mode of progression which enables it to attain ten or twelve miles an hour, which Mr. Beale believes to be its greatest velocity; 'lob-tailing,' or lashing the water with its tail; and the vivid descriptions of the dangers and hair-breadth escapes attending its capture, we must refer to Mr. Beale's book, which every one who is anxious for information on this subject should read. Other information on the whale fishery is contained in the article FISHERIES, vol. x., p. 288.

Geographical Distribution.—Very wide: the species has been seen in almost all seas, but it is now principally found in the Southern Ocean, on the coasts of America, Japan, New Guinea, Timor, &c. Cachalots have occasionally been stranded in the British Islands, as in the Frith of Forth [*sic*], the Orkneys, &c. . . .

The *Common Whale*, *Greenland Whale*, or *Great Mysticete*, appears to be the μυσικητος of Aristotle, (Hist. Anim. III. 12.); it is the *Baleine Franche* and *Baleine Ordinaire* of the French; *Gronlands Walfisk* of the Swedes; and *Morfil Cyffredin* of the antient [*sic*] British.

Habits, Food, &c.—This species seems to hear acutely any noise made in the water, such as splashing, &c. in calm weather: but a sound produced in the air, a loud shout for instance, when the whale is only at the distance of a ship's length, is disregarded. The usual rate of swimming seldom exceeds four miles an hour, but they will descend when harpooned at a velocity of seven or eight miles an hour, and one of these whales when alarmed can sink in five or six seconds far beyond the reach of a human enemy. The Mysticete seldom remains at the surface to breathe longer than two minutes, during which period it blows eight or nine times. It then descends for five or ten minutes; sometimes, when on its feed, for fifteen or twenty minutes. Though Mr. Scoresby states that it has no voice, it makes, he observes, a loud noise in blowing. The spout is ejected some yards high, and has the appearance of a puff of smoke at a distance. They blow strongest, densest, and loudest, when alarmed, or after a long stay under water.

A very considerable portion of the feeding grounds is occupied by what is termed "green water," which swarms with minute life, and has been carefully examined and described by Mr. Scoresby. The smallness of the gullet is only fitted for swallowing small animals, such as the *Clio borealis*, numerous specimens of which, the *Whale's Food* of the Greenland Whalers, will be found in the preparation No. 323 A of the Physiological series of the Museum of the Royal College of Surgeons in London. This small mollusk is said to constitute the chief support of the Mysticete, and the structure and disposition of the whalebone plates, illustrated in the preceding preparations, explain how these or any other small species of animal are retained in the capacious mouth of their devourer, while the water taken in along with them drains through the interstices of the plates. When the Mysticete feeds, it swims rapidly below the surface with open jaws; a stream of water enters them, and with it myriads of small marine animals; the water finds an outlet at the sides, but the thick internal hairy apparatus of the whalebone does not permit one of those animals to escape.

Nine or ten months is supposed to be the period of utero-gestation, and the mother is so attached to her young one, or "sucker," as it is termed, that it is often struck as a snare to the affectionate parent, for she will not leave it, and falls a victim to her maternal love. Mr. Scoresby relates instances of this kind which cannot be perused, much less witnessed, without great pain by any person of ordinary humanity. Such a mode of capture seems hardly justifiable, whilst it must be ruinous to future prospects.

This species is generally found alone or in pairs, excepting when many individuals are attracted to some abundant feeding ground, or to a desired locality, such as the vicinity of ice-bergs.

Utility to Man.—To the Esquimaux and the Greenlander this species is all in all. They eat the flesh and fat with indescribable relish. The membranes of the abdomen serve them for clothing, and the thin transparent peritoneum admits light through the windows of their huts whilst it keeps out the weather. The bones are made into props for their tents, or aid in the formation of their boats, and supply them with harpoons and spears for the capture of the seal, and greater sea-birds. The sinews divided into filaments are used as thread for sewing their dress, &c. Some have stated that pickled and boiled blubber is palatable, and that the tail, first parboiled and the fried, is agreeable eating. The flesh of the young whale is said to be by no means indifferent food. To civilized nations, the oil made from its fat or blubber, and the whalebone, have long made it a great commercial object. [FISHERIES, vol. x., p. 288.]

To Mr. Scoresby's well-known book we must refer for further interesting and amusing particulars. The history of the *Balænopteræ*, or *Rorquals*, some of which grow to eighty or a hundred feet in length, may be found in the works of Lacépède, and of Baron Cuvier and M. F. Cuvier.

Notes

1. London, John Van Voorst, 1839, 8vo.
2. 'Narrative of a Whaling Voyage round the Globe,' 2 vols. 8vo., Bentley, 1840. Many interesting facts in natural history are recorded in this entertaining book.

From Henry Trumbull, *Life and Remarkable Adventures of Israel R. Potter*

(Providence: Henry Trumbull, 1824)

INTRODUCTION

Trumbull (1781–1843) was a marginally productive writer from Stonington, Connecticut, the home also of sealing captain Nathaniel B. Palmer and, in fact, the center of the American sealing industry. Trumbull primarily wrote popular but often inaccurate adaptations of colonial history.

When Melville adapted Trumbull's narrative, he promoted Israel to harpooner (as he did for himself in his letter to publisher Richard Bentley in June 1850). Israel's second whaling voyage must have taken place around 1773, antedating by 15 years the first recorded Pacific whaling voyage, that of Samuel Enderby's *Emilia* departing in 1788. Three years is an appropriate time for such a voyage, and the addition must have been a sophistication by Trumbull. Surely, a scholar of whaling sources like Melville would have caught the error—unless he adapted this part of Trumbull's narrative *before* his intensive cetological studies. Melville also makes Israel an expert in "darting the whale-lance," another mixture of whaling terminology that would be hard to explain if the author had just refreshed his memory with a variety of whaling authorities. Cf. "Chapter 62 The Dart" in *Moby-Dick.* Melville encountered the volume before leaving for England in the autumn of 1849. Sealts #407; Bercaw #721. Modern edition: in NN *Israel Potter* (1982).

PRIMARY SOURCE

After remaining at home one month, I applied for and procured a birth at Providence, on board the Sloop ———, Capt. Fuller, bound to Grenada—having completed her loading (which consisted of stone lime, hoops, staves, &c.) we set sail with a favorable wind, and nothing worthy of note occurred until the 15th day from that on which we left Providence, when the sloop was discovered to be on fire, by a smoke issuing from her hold—the hatches were immediately raised, but as it was discovered that the fire was caused by water communicating with the lime, it was deemed useless to make any attempts to extinguish it—orders were immediately thereupon given by the captain to hoist out the long boat, which was found in such a leaky condition as to require constant bailing to keep her afloat: we had only time to put on board a small quantity of bread, a firkin of butter and a ten gallon keg of water, when we embarked, eight in number, to trust ourselves to the mercy of the waves, in a leaky boat and many leagues from land. As our provision was but small in quantity, and it being uncertain how long we might remain in our perilous situation, it was proposed by the captain soon after leaving the sloop, that we should put ourselves on an allowance of one biscuit and half a pint of water per day, for each man, which was readily agreed to by all on board—in ten minutes after leaving the sloop she was in a complete blaze, and presented an awful spectacle. With a piece of the flying-jib, which had been fortunately thrown into the boat, we made shift to erect a sail, and proceeded in a south west direction in hopes to reach the spanish maine, if not so fortunate as to fall in with some vessel in our course—which, by the interposition of kind providence in our favour, actually took place the second day after leaving the sloop—we were discovered and picked up by a Dutch ship bound from Eustatia to Holland, and from the captain and crew met with a humane reception, and were supplied with every necessary that the ship afforded—we continued on board one week when we fell in with an American sloop bound from Piscataqua to Antigua, which received us all on board and conveyed us in safety to the port of her destination. At Antigua I got a birth on board an American brig bound to Porto Rico, and from thence to Eustatia. At Eustatia I received my discharge and entered on board a Ship belonging to Nantucket, and bound on a whaling voyage, which proved an uncommonly short and successful one—we returned to Nantucket full of oil after an absence of the ship from that port of only 16 months. After my discharge I continued about one month on the island, and then took passage for Providence, and from thence went to Cranston, once more to visit my friends, with whom I continued three weeks, and then returned to Nantucket. From Nantucket

I made another whaling voyage to the South Seas and after an absence of three years, (in which time I experienced almost all the hardships and deprivations peculiar to Whalemen in long voyages) I succeeded by the blessings of providence in reaching once more my native home, perfectly sick of the sea, and willing to return to the bush and exchange a mariner's life for one less hazardous and fatiguing.

C. HAMILTON SMITH, "WHALE," IN JOHN KITTO, *CYCLOPÆDIA OF BIBLICAL LITERATURE*

VOL. 2 (1845)

INTRODUCTION

John Kitto (1804–1854), although never healthy, overcame profound childhood financial and physical handicaps to become an energetic and prolific evangelical Biblical scholar, writing for—among others—Charles Knight's *Penny Magazine.* Charles Hamilton Smith (1776–1859) was a polymathic adventurer and artist. The collaboration seems unlikely, but seems to have been an effective one, as Kitto managed a large and diverse stable of contributors to his *Cyclopædia.* An abridged one-volume edition was published in 1849, but Melville seems to have used the longer entry of the original two-volume edition. The absence of scholarly attention to Melville's use of Kitto before *Moby-Dick* should not automatically be taken as proof that he had not acquired the work. See Ilana Pardes, "Remapping Jonah's Voyage: Melville's *Moby-Dick* and Kitto's *Cyclopedia of Biblical Literature,*" *Comparative Literature* 57:2 (Spring 2005), 135–157. Bercaw #421.

PRIMARY SOURCE

WHALE (תן *than*, and תנין *thannin;* Sept. and Matt. xii.40, κῆτος), occurs in several places of the Old Testament, and once in the New Testament. In the passages where scales and feet are mentioned as belonging to *than*, commentators have shown that the crocodile is intended, which then is synonymous with the leviathan; and they have endeavoured also to demonstrate, where *thannin* draw the dugs to suckle their young, that seals are meant, although cetacea nourish theirs in a similar manner. It may be doubted whether, in most of the cases, the poetical diction points absolutely to any specific animal, particularly as there is more force and grandeur in a generalized and collective image of the huge monsters of the deep, not inappropriately so called, than in the restriction to any one species, since all are in Gen. i.26 made collectively subservient to the supremacy of man. But criticism is still more inappropriate when, not contented with pointing to some assumed species, it attempts to rationalise miraculous events by such arguments; as in the case of Jonah, where the fact of whales having a small gullet, and not being found in the Mediterranean, is adduced to prove that the huge fish דג *dag* was not a cetacean, but a shark! Now, if the text be literally taken, the transaction is plainly miraculous, and no longer within the sphere of zoological discussion; and if it be allegorical, as some, we think, erroneously assume, then, whether the prophet was saved by means of a kind of boat called *dagh*, or it be a mystical account of initiation where the neophite was detained three days in an ark or boat, figuratively denominated a fish, or Celtic *avanc*, the transaction is equally indeterminate; and it assuredly would be derogating from the high dignity of the prophet's mission, to convert the event into a mere escape, by boat, or into a pagan legend such as Hercules, Bacchus, Jamsheed, and other deified heroes of the remotest antiquity, are fabled to have undergone, and which all the ancient mysteries, including the Druidical, symbolized. It may be observed, besides, of cetaceous animals, that though less frequent in the Mediterranean than in the ocean, they are far from being unknown there. Joppa, now Jaffa, the very place whence Jonah set sail, displayed for ages in one of its pagan temples huge bones of a species of whale, which the legends of the place pretended were those of the dragon monster slain by Perseus, as represented in the Arkite mythus of that hero and Andromeda and which remained in that spot till the conquering Romans carried them in triumph to the great city. Procopius mentions a huge sea-monster in the Propontis, taken during his præfecture of Constantinople, in the 36th year of Justinian, (A.D. 562), after having destroyed vessels at certain intervals for more than fifty years. Rondoletius enumerates several whales stranded or taken on the coasts of the Mediterranean: these were most likely all *orcas*, *physeters*, or *campedolios*, i.e. toothed whales, as large and more fierce than the *mysticetes*, which have balein in the mouth, and at present very rarely make their way farther south than the Bay of Biscay; though in early times it is probable they visited the Mediterranean, since the present writer has seen them within the tropics. In the Syrian seas, the Belgian pilgrom Lavaers, on his passage from Malta to Palestine, incidentally mentions a 'Tonyuvisch,' which he further denominates an 'oil-fish,' longer than the vessel, leisurely swimming along, and which the seamen said prognosticated bad weather. On the island of Zerbi, close to the African coast, the late Commander Davies, R.N., found the bones of a cachalot on the beach. Shaw mentions an orca more than sixty feet in length, stranded at Algiers; and the late Admiral Ross Donelly saw one in the Mediterranean near the island of Albaran. There are, besides, numerous sharks of the largest species in the seas of the Levant, and also in the Arabian Gulf and Red Sea, as well as cetacea, of which *balæna bitan* is the largest in those seas, and two species of *halicore* or *dugong*, which are herbivorous animals, intermediate between whales and seals.—C.H.S.

From William Scoresby, *Account of the Arctic Regions, with a History and Description of the Northern Whale Fishery*

(Edinburgh: Constable, 1820)

INTRODUCTION

William Scoresby (1789–1857) was the son of a successful Yorkshire whaleman of the same name. He learned the craft of whaling while accompanying his father on voyages to the Arctic Ocean, the haunt of the Right or Greenland Whale. He also exercised a strong interest in both Christianity and natural history, which he did not find mutually exclusive. He became a friend of the greatest scientist-explorer of the age, Sir Joseph Banks (1743–1820), and was readily elected to scientific organizations when his gift for observation became known.

Although presumably the first whaling source garnered by Melville after Dana's 1850 letter (he borrowed the volumes on 29 April), Scoresby's two-volume work offered little Melville could use directly in an account of the Southern Whale Fishery. Nevertheless, Melville found ample occasion to make fun of Scoresby's indefatigable and detailed pursuit of knowledge and found such passages as the first one below useful to corroborate the migrations of whales. From other passages, Melville seems to have straightforwardly adopted Scoresby's first-person authoritative voice, even when its point of view was that of a harpooner or officer—neither a rank Ishmael possessed in *Moby-Dick*. Melville was also familiar with Scoresby's *Journal of a Voyage to the Northern Whale Fishery* (1823). See Vincent (1849). Sealts # 450; Bercaw #616. Modern edition: Newton Abbot: David & Charles, 1969.]

PRIMARY SOURCE

General Remarks Indicating the Existence of a Sea Communication between the Atlantic and Pacific Oceans, by the North

. . . another argument which goes still farther to support the opinion of the existence of the communication in question, is the fact of whales which have been harpooned in the Greenland seas, having been found in the Pacific Ocean; and whales with stone lances sticking in their fat, (a kind of weapon used by no nation now known,) having been caught both in the sea of Spitzbergen, and in Davis' Strait. The following are some of the authorities for this fact, which, of all other arguments yet offered in favour of a trans-polar passage, seems to me to be the most satisfactory.

A Dutch East India captain, of the name of Jacob Cool, of Sardam, who had been several times at Greenland, and was of course well acquainted with the nature of the apparatus used in the whale-fishery, was informed by the Fischal Zeeman of India, that in the sea of Tartary there was a whale taken, in the back of which was sticking a Dutch harpoon, marked with the letters W. B. This curious circumstance was communicated to Peter Jansz Vischer, probably a Greenland whaler, who discovered that the harpoon in question had belonged to William Bastiaanz, Admiral of the Dutch Greenland fleet, and had been struck into the whale in the Spitzbergen sea.[1]

Muller refers to a similar circumstance when recording the first discovery by sea, of the peninsula of Kamtchatka by the Russians, in the year 1716. The crew of the discovery vessel having wintered on the western coast of Kamtchatka, he informs us, that during their stay there, "the sea cast upon the shore a whale that had in its body a harpoon of European workmanship, marked with Roman letters."[2] Another account of the same nature, given by Hendrick Hamel, in his "Unfortunate Voyage of the yacht Sparwer, in the year 1653," and published in the "Recuéil des Voyages," corroborates the testimony of Muller. Hamel, in his narrative of the loss of this vessel on the Island of Quelpaert, observes, that "in the sea to the north-east of Korea, they take every year a great number of whales, in some of which are found harpoons (or striking-irons) of the French and Dutch, who practise the whale-fishery at the extremities of Europe; whence we infer (he continues) that there is surely a passage between Korea and Japan, which communicates to the Strait of Waigatz,"[3] separating Nova Zembla from the Continent of Europe.

Other circumstances can be adduced to the same effect. The master of the Volunteer whaler of Whitby, when near the coast of Spitzbergen, July 19. 1813, shewed me part of a lance which had been taken out of the fat of a whale killed by his crew a few weeks before. It was formed of a hard grey stone, of a flinty appearance, about three inches long, two broad, and two-tenths thick. Two holes were pierced in one end of it, by which, it appeared the stock or handle had been secured. It was completely embedded in the blubber, and the wound was quite healed. A small white scar on the skin of the whale, alone marked the place where the lance had entered. In the year 1812, the crew of a Hull fisher (the Aurora) met with a whale in the same region having a harpoon made of bone, sticking in its back; and a few years ago a lance of stone, somewhat like the one above mentioned, fixed to a piece of bone, forming a socket for the stock, was likewise found in a whale by the people of another Greenlandman of Hull: this stone-lance is now deposited in the interesting collection of natural rarities belonging to Mr Hornsea of Scarbrough. To these facts we might add many of a similar kind, together with others of whales struck in Davis' Straits having been killed near Spitzbergen, and *vice versa*; but the above will be sufficient for affording a strong confirmation of the opinion, that a sea communication between the Atlantic and Pacific Oceans by the North, does exist. For, with regard to the stone-lances and bone-harpoons found in the bodies of whales, it may be remarked, that as the Esquimaux of Davis' Straits and Hudson's Bay, have now, from their long intercourse with Europeans, become well supplied with weapons calculated for the capture of the whale made of iron, these instruments of stone and bone, so much inferior, must have been used by some other persons who have not yet had intercourse with the civilized world; but as they are precisely the kind of weapons which were in common use among the Esquimaux a century ago, it is probable that the instruments alluded to were struck by some tribe of the same nation, inhabiting the shores of the frozen ocean, on the northern face of the American Continent, yet unexplored. If so, these facts go far towards establishing the existence of a communication between the Spitzbergen sea and the Pacific Ocean.

Notes

1. Beschryving der Walvisvangst, vol. ii. p. 38
2. Muller's Voyages from Asia to America; Jeffrey's Translation, p. 42.
3. Quarterly Review, No. xxxv. p. 217.

A Description of Animals, of the Cetaceous Kind frequenting the Greenland Sea

Balæna Mysticetus: The Common Whale, or Greenland Whale

This valuable and interesting animal, generally called *The Whale* by way of eminence, is the object of our most important commerce to the Polar Seas,—is productive of more oil than any other of the Cetacea, and, being less active, slower in its motion, and more timid than any other of the kind, of similar or nearly similar magnitude, is more easily captured.

Large as the size of the whale certainly is, it has been much over-rated;—for such is the avidity with which the human mind receives communications of the marvellous, and such the interest attached to those researches which describe any remote and extraordinary production of nature, that the judgment of the traveller receives a bias, which, in cases of doubt, induces him to fix upon that extreme point in his opinion which is calculated to afford the greatest surprise and interest. Hence, if he perceives an animal remarkable for its minuteness, he is inclined to compare it with something still more minute;—if remarkable for its bigness, with something fully larger. When the animal inhabits an element where he cannot examine it, or is seen under any circumstances which prevent the possibility of his determining its dimensions, his decision will certainly be in that extreme which excites the most interest. Thus a mistake in the size of the whale would be easily made. And there is every probability of such an error having been committed two or three centuries back, from which period some of our present dimensions have been derived, when we know that whales were usually viewed with superstitious dread, and their magnitude and powers, in consequence, highly exaggerated. Besides, errors of this kind having a tendency to increase rather than to correct one another, from the circumstance of each writer on the subject being influenced by a similar bias, the most gross and extravagant results are at length obtained.

Thus authors, we find, of the first respectability in the present day, give a length of 80 to 100 feet, or upwards, to the Mysticetus, and remark, with unqualified assertion, that when the captures were less frequent, and the animals had sufficient time to attain their full growth, specimens were found of 150 to 200 feet in length, or even longer; and some ancient naturalists, indeed, have gone so far as to assert, that whales had been seen of above 900 feet in length. But whales, in the present day, are by no means so bulky. Of 322 individuals, in the capture of which I have been personally concerned, no one, I believe, exceeded 60 feet in length; and the largest I ever measured, was 58 feet from one extremity to the other, being one of the longest, to appearance, which I ever saw. An uncommon whale that was caught near Spitzbergen, about 20 years ago, the whalebone of which measured almost 15 feet, was not, I understand, so much as 70 feet in length; and the longest actual measurement that I have met with, or heard of, is given by Sir Charles Giesecké, who informs us, that, in the spring of 1813, a whale was killed at Godhavn, of the length of 67 feet. These, however, are very uncommon instances. I therefore conceive, that 60 feet may be considered as the size of the larger animals of this species, and 65 feet in length as a magnitude which very rarely occurs. Yet I believe that whales now occur of as large dimensions, as at any former period since the commencement of the whale-fishery. This point I endeavoured to prove, from various historical records, in a paper read before the Wernerian Society, on the 19th of December 1818, and since inserted in the "Edinburgh Philosophical Journal," No. I. p. 83.

In this paper, I brought forward the authorities of Zorgdrager, the writer of an account of the whale fishery, and one of the early superintendants [*sic*] of the Dutch northern fisheries, together with opinions or remarks of Captains Anderson, Gray, Heley, and others, who were among the earliest of the English whalers, which satisfactorily prove, that the average and largest produce of a whale in oil, was not greater, near two hundred years ago, than it is at the present time; and to these are added, the testimonies of Captains Jenkinson and Edge, as to the length of the whale, which likewise corresponds, pretty nearly, with the measurements I have myself made.

Jenkinson, in his voyage to Russia, performed in 1557, saw a number of whales, some of which, by estimation, were 60 feet long, and are described as being "very monstrous." Edge, who was one of the Russia Company's chief and earliest whalefishers, having been ten years to Spitzbergen, prior to the year 1625, calls the whale "a sea beaste of huge bigness, about 65 foot long and 35 foot thick," having whalebone 10 or 11 feet long, (a common size at present,) and yielding about 100 hogsheads of oil; and, in a descriptive plate accompanying Captain Edge's paper on the fishery, published by Purchas in 1625, is a sketch of a whale, with this remark subjoined,— "A whale is ordinarily about 60 foot long." Hence I conceive we may satisfactorily conclude, that whales of as large size are found now, as at any former period since the Spitzbergen fishery was discovered; and I may also remark, that where any respectable authority affords actual measurements exceeding 70 feet, it will always be found that the specimen referred to, was not one of the Mysticetus kind, but of the B, Physalis, or the B. Musculus, animals which considerably exceed in length any of the common whales that I have either heard of, or met with.

When fully grown, therefore, the length of the whale may be stated as varying from 50 to 65, and rarely, if ever, reaching 70 feet; and its greatest circumference from 30 to 40 feet. . . .

Anecdotes Illustrative of the Dangers of the Whale-Fishery

Those employed in the occupation of killing whales, are, when actually engaged, exposed to danger from three sources, viz. from the ice, from the climate, and from the whales themselves. . . .

The most extensive source of danger to the whale-fisher, when actively engaged in his occupation, arises from the object of his pursuit.

Excepting when it has young under its protection, the whale generally exhibits remarkable timidity of character. A bird perching on its back alarms it; hence, the greater part of the accidents which happen in the course of its capture, must be attributed to adventitious circumstances on the part of the whale, or to mismanagement or fool-hardiness on the part of the fishers.

In this employment, the fisher is liable to receive concussions from oars forcibly struck by the fish, or direct blows from its fins or tail;—he is liable to accidents, from getting entangled by the lines, or from the boat being drawn under water by the fish, through the medium of the lines;—and he is in danger of being thrown overboard by the heeling or jerking of the boat, or more particularly from the boat being stove, upset, sunk, or projected into the air, by the force of a blow from the whale.

Contusions by blows from the whale.—Several instances have occurred, in which serious accidents have been occasioned by blows from the whale either directly struck, or impressed through the medium of oars. Harpooners have been struck dead by a single blow of a whale's tail, and other fishers have received dangerous wounds. But as I am not in possession of so many of the particulars of these circumstances as to be able to give them in the narrative form, this description of accidents must pass without any particular illustration.

Injuries sustained by Entanglements of the Lines.—One of the crew of the John of Greenock, who was in a fast-boat, in the fishery of 1818, unfortunately slipped his foot through a coil of line in the act of running out, which drew him forward to the boat's stem, and separated his foot by the ancle [*sic*]. He was conveyed by the first boat to the ship, where the assistance of several surgeons being procured, the lower part of the leg was cut off. After this, the poor fellow having received the most unremitting attention from Captain Jackson, with the best sustenance and accommodation the ship could afford, was restored to health, and his wound in a great measure healed, before the conclusion of the voyage. It is worthy of being remarked, that the captain and crew of the John subscribed upwards of 24£ for his relief, which was increased by the owners of the ship, and others, on arrival, to about 37£. This sum was placed in the "Provident Bank," at Greenock, from whence he was permitted to draw it, after the rate of 7*s. per* week.

A harpooner belonging to the Henrietta of Whitby, when engaged in lancing a whale, into which he had previously struck a harpoon, incautiously cast a little line under his feet that he had just hauled into his boat, after it had been drawn out by the fish. A painful stroke of his lance induced the whale to dart suddenly downward, his line to run out from beneath his feet, and in an instant caught him by a turn round his body. He had but just time to cry out, "Clear away the line,"—"O dear!" when he was almost cut asunder, dragged overboard, and never seen afterwards. The line was cut at the moment, but without avail. The fish descended a considerable depth, and died; from whence it was drawn to the surface by the lines connected with it, and secured.

Fishers thrown Overboard, by the jerking or sudden heeling of the Boats, in consequence of Blows from Whales.—On the 3d of June 1811, a boat from the ship Resolution, commanded at the time by myself, put off in pursuit of a whale, and was rowed upon its back. At the moment that it was harpooned, it struck the side of the boat a violent blow with its tail, the shock of which threw the boat-steerer some distance into the water. A repetition of the blow projected the harpooner and line-manager in a similar way, and completely drenched the part of the crew remaining in the boat, with the sprays. One of the men regained the boat, but as the fish immediately sunk, and drew the boat away from the place, his two companions in misfortune were soon left far beyond the reach of assistance. The harpooner, though a practised swimmer, felt himself so bruised and enervated by a blow he had received on the chest, that he was totally incapacitated from giving the least support to his fellow sufferer. The ship being happily near, a boat which had been lowered on the first alarm, arrived to their succour, at the moment when the line-manager, who was unacquainted with the art of swimming, was on the point of sinking, to rise no more. Both the line-manager and harpooner were preserved; and the fish, after a few hours close pursuit, was subdued.

A large whale, harpooned by a boat belonging to the same ship, became the subject of a general chase on the 23d of June 1809. Being myself in the first boat which approached the fish, I struck my harpoon at arm's length, by which we fortunately evaded a blow that appeared to be aimed at the boat. Another boat was then advanced, and another harpoon was struck, but not with the same result; for the stroke was immediately returned by a tremendous blow from the fish's tail. The boat was sunk by the shock; and, at the same time, whirled round with such velocity, that the boat-steerer was precipitated into the water, on the side next to the fish, and was accidentally carried down to a considerable depth by its tail. After a minute or so, he arose to the surface of the water and was taken up, along with his companions into my boat. A similar attack was made on the next boat which came up; but the harpooner being warned of the prior conduct of the fish, used such precautions, that the blow, though equal in strength, took effect only in an inferior degree. The boat was slightly stove. The activity and skill of the lancers soon overcame this designing whale, accomplished its capture, and added its produce to the cargo of the ship. Such intentional mischief on the part of a whale, it must be observed, is an occurrence which is somewhat rare.

Again. While the same ship, (Resolution), navigated an open lake of water, in the 81st degree of north latitude, during a keen frost and strong north wind, on the 2d of June 1806, a whale appeared, and a boat put off in pursuit. On its second visit to the surface of the sea, it was harpooned. A convulsive heave of the tail, which succeeded the wound, struck the boat at the stern; and by its reaction, projected the boat-steerer overboard. As the line in a moment dragged the boat beyond his reach, The crew threw some of their oars towards him for his support, one of which he fortunately seized. The ship and boats being at a considerable distance, and the fast-boat being rapidly drawn away from him, the harpooner cut the line, with the view of rescuing him from his dangerous situation. But no sooner was this act performed, than to their extreme mortification they discovered, that in consequence of some oars being thrown towards their floating comrade, and others being broken or unshipped by the blow from the fish, one oar only remained; with which, owing to the force of the wind, they tried in vain to approach him. A considerable period elapsed, before any boat from the ship could afford him assistance, though the men strained every nerve for the purpose. At length, when they reached him, he was found with his arms stretched over an oar, almost deprived of sensation. On his arrival at the ship, he was in a deplorable condition. His clothes were frozen like mail, and his hair constituted a helmet of ice. He was immediately conveyed into the cabin, his clothes taken off, his limbs and body dried and well rubbed, and a cordial administered to him which he drank. A dry shirt and stockings were then put upon him, and he was laid in the captain's bed. After a few hours sleep he awoke, and appeared considerably restored, but complained of a painful sensation of cold. He was, therefore, removed to his own *birth*, and one of his messmates ordered to lie on each side of him, whereby the diminished circulation of the blood was accelerated, and the animal heat restored. The shock on his constitution, however, was greater than was anticipated. He recovered in the course of a few days, so as to be able to engage in his ordinary pursuits; but many months elapsed, before his countenance exhibited its wonted appearance of health.

Boats Sunk, Stove, or Upset, by blows from Whales.—The Aimwell of Whitby, while cruising the Greenland seas, in the year 1810, had boats in chase of whales on the 26th of May. One of them was harpooned. But instead of sinking immediately on receiving the wound, as is the most usual manner of the whale, this individual only dived for a moment, and then rose again beneath the boat, struck it in the most vicious manner with its fins and tail, stove it, upset it, and then disappeared. The crew, seven in number, got on the bottom of the boat; but the unequal action of the lines, which for some time remained entangled with the boat, rolled it occasionally over, and thus plunged the crew repeatedly into the water. Four of them, after each immersion, recovered themselves and clung to the boat; but the other three, one of whom was the only person acquainted with the art of swimming, were drowned before assistance could arrive. The four men on the boat being rescued and conveyed to the ship, the attack on the whale was continued, and two more harpoons struck. But the whale irritated, instead of being enervated by his wounds, recommenced its furious conduct. The sea was in foam. Its tail and fins were in awful play; and in a short time, harpoon after harpoon drew out, the fish was loosened from its entanglements and escaped.

In the fishery of 1812, the Henrietta of Whitby suffered a similar loss. A fish which was struck very near the ship, by a blow of its tail, stove a small hole in the boat's bow. Every individual shrinking from the side on which the blow was impressed, aided the influence of the stroke, and upset the boat. They all clung to it while it was bottom up; but the line having got entangled among the thwarts, suddenly drew the boat under water, and with it part of the crew. Excessive anxiety among the people in the ship, occasioned delay in sending assistance; so that when the first boat arrived at the spot, two survivors only out of six men were found.

During a fresh gale of wind in the season of 1809, one of the Resolution's harpooners struck a sucking whale. Its mother being near, all the other boats were disposed around, with the hope of entangling it. The old whale pursued a circular route round its cub, and was followed by the boats; but its velocity was so considerable, that they were unable to keep pace with it. Being in the capacity of harpooner on this occasion myself, I proceeded to the chase, after having carefully marked the proceedings of the fish. I selected a situation, in which I conceived the whale would make its appearance, and was in the act of directing my crew to cease rowing, when a terrible blow was struck on the boat. The whale I never saw, but the effect of the blow was too important to be overlooked. About 15 square feet of the bottom of the boat were driven in;

it filled, sunk, and upset in a moment. Assistance was providentially at hand, so that we were all taken up without injury, after being but a few minutes in the water. The whale escaped; the boat's lines fell out and were lost, but the boat was recovered.

A remarkable instance of the power which the whale possesses in its tail, was exhibited within my own observation, in the year 1807. On the 29th of May, a whale was harpooned by an officer belonging to the Resolution. It descended a considerable depth; and, on its re-appearance, evinced an uncommon degree of irritation. It made such a display of its fins and tail, that few of the crew were hardy enough to approach it. The captain, (my Father,) observing their timidity, called a boat, and himself struck a second harpoon. Another boat immediately followed, and unfortunately advanced too far. The tail was again reared into the air, in a terrific attitude,—the impending blow was evident,—the harpooner, who was directly underneath, leaped overboard,—and the next moment the threatened stroke was impressed on the centre of the boat, which buried it in the water. Happily no one was injured. The harpooner who leaped overboard, escaped certain death by the act,—the tail having struck the very spot on which he stood. The effects of the blow were astonishing. The keel was broken,—the gunwales, and every plank, excepting two, were cut through,—and it was evident that the boat would have been completely divided, had not the tail struck directly upon a coil of lines. The boat was rendered useless.

Instances of disasters of this kind, occasioned by blows from the whale, could be adduced in great numbers,—cases of boats being destroyed by a single stroke of the tail, are not unknown,—instances of boats being stove or upset, and their crews wholly or in part drowned, are not unfrequent,—and several cases of whales having made a regular attack upon every boat which came near them, dashed some in pieces, and killed or drowned some of the people in them, have occurred within a few years, even under my direct observation.

Boats, together with their Crews and Apparatus, projected into the Air.— The Dutch ship Gort-Moolen, commanded by Cornelius Gerard Ouwekaas, with a cargo of seven fish, was anchored in Greenland in the year 1660. The captain, perceiving a whale a-head of his ship, beckoned his attendants, and threw himself into a boat. He was the first to approach the whale; and was fortunate enough to harpoon it before the arrival of the second boat, which was on the advance. Jacques Vienkes, who had the direction of it, joined his captain immediately afterwards, and prepared to make a second attack on the fish, when it should remount again to the surface. At the moment of its ascension, the boat of Vienkes happening unfortunately to be perpendicularly above it, was suddenly and forcibly lifted up by a stroke of the head of the whale, that it was dashed to pieces before the harpooner could discharge his weapon. Vienkes flew along with the pieces of the boat, and fell upon the back of the animal. This intrepid seaman, who still retained his weapon in his grasp, harpooned the whale on which he stood, and, by means of the harpoon and the line, which he never abandoned, he steadied himself firmly upon the fish, notwithstanding his hazardous situation, and regardless of a considerable wound that he received in his leg, in his fall along with the fragments of the boat. All the efforts of the other boats to approach the whale, and deliver the harpooner, were futile. The captain, not seeing any other method of saving his unfortunate companion, called to him to cut it with his knife, and betake himself to swimming. Vienkes, embarrassed and disconcerted as he was, tried in vain to follow this counsel. His knife was in the pocket of his drawers; and, being unable to support himself with one hand, he could not get it out. The whale, mean while, continued advancing along the surface of the water with great rapidity, but fortunately never attempted to dive. While his comrades despaired of his life, the harpoon by which he held, at length disengaged itself from the body of the whale. Vienkes being then liberated, did not fail to take advantage of this circumstance; he cast himself into the sea, and, by swimming, endeavoured to regain the boats which continued the pursuit of the whale. When his shipmates perceived him struggling with the waves, they redoubled their exertions. They reached him just as his strength was exhausted, and had the happiness of rescuing this adventurous harpooner from his perilous situation.[4]

In one of my earliest voyages to the whale-fishery, I observed a circumstance which excited my highest astonishment. One of our harpooners had struck a whale, it dived, and all the assisting boats had collected round the fast-boat, before it arose to the surface. The first boat which approached it advanced incautiously upon it. It rose with unexpected violence beneath the boat, and projected it and all its crew, to the height of some yards in the air. It fell on its side, upset, and cast all the men into the water. One man received a severe blow in his fall, and appeared to be dangerously injured; but soon after his arrival on board of the ship, he recovered from the effects of the accident. The rest of the boat's crew escaped without any hurt.

Captain Lyons of the Raith of Leith, while prosecuting the whale-fishery on the Labrador coast, in the season of 1802, discovered a large whale at a short distance from the ship. Four boats were dispatched in pursuit, and two of them succeeded in approaching it so closely together, that two harpoons were struck at the same moment. The fish descended a few fathoms in the direction of another of the boats, which was on the advance, rose accidentally beneath it, struck it with its head, and threw the boat, men and apparatus, about fifteen feet into the air. It was inverted by the stroke, and fell into the water with its keel upwards. All the people were picked up alive by the fourth boat, which was just at hand, excepting one man, who having got entangled in the boat, fell beneath it, and was unfortunately drowned. The fish was soon afterward killed. . . .

Note

4. I give this anecdote on the authority of the author of the Histoire de Péches, who translated it from the Dutch. Part of the story bears the marks of truth; but some of it, it must be acknowledged, borders on the marvellous.

From Thomas Beale, *The Natural History of the Sperm Whale*

(London: Van Voorst, 1839)

INTRODUCTION

Thomas Beale (1807–1849) left England on a whaling voyage to the South Seas beginning in 1830. He served as surgeon on two vessels, the *Kent* and the *Sarah and Elizabeth*, before returning in 1833. His book has long been regarded as the most important of Melville's cetological authorities, in part because of Melville's heavy annotations in his copy of Beale's book and in part because of the widespread publication of excerpts from it. Beale himself had published an article on the Sperm Whale in the *Quarterly Review* and a preliminary pamphlet without the narrative portion, titled *A Few Observations on the Natural History of the Sperm Whale* (1835). Although Beale was awarded a silver medal by the Eclectic Society in 1834 for his paper on the Sperm Whale, biographical information apart from his book remains scant. Melville's copy of Beale arrived from England by July 10, 1850. See Vincent (1949) and Olsen-Smith (2010). Sealts #52; Bercaw #52. Modern edition: Holland Press, 1973.

PRIMARY SOURCE

External form and Peculiarities of the Sperm Whale

Before proceeding to the account of the habits of the sperm whale, I have thought that it might be interesting to prefix a short description of its external

Fig. 1.

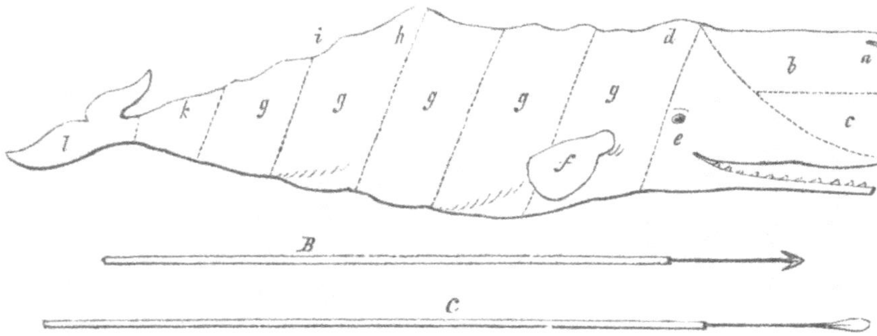

Fig. 1 represents the outline of the entire form.
Fig. 2, that of the anterior aspect of the head.

Fig. 2.

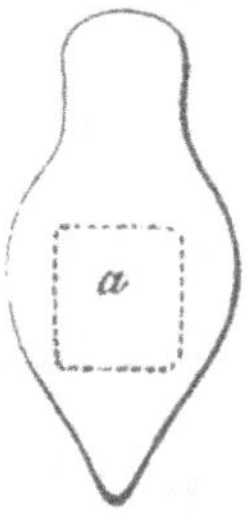

Fig. 1—*a*, the nostril or spout-hole; *b*, the situation of the case; *c*, the junk; *d*, the bunch of the neck; *e*, the eye; *f*, the fin; *g*, the spiral strips, or blanket pieces; *h*, the hump; *i*, the ridge; *k*, the small; *l*, the tail or flukes; *B*, a harpoon; *C*, a lance.

Fig. 2—*a*, the lines forming the square are intended to represent the flat anterior part of the head.

Thomas Beale, *The Natural History of the Sperm Whale* (London: Van Voorst, 1839)

form, and some anatomical points in its conformation. By reference to the prefixed engravings, the following description will be much more readily understood:

The head of the sperm whale presents in front a very thick blunt extremity, called the snout or nose, and constitutes about one-third of the whole length of the animal–at its junction with the body is a large protuberance on the back, called by the whalers the "bunch of the neck;" immediately behind this, or at what might be termed the shoulder, is the thickest part of the body, which from this point gradually tapers off to the tail, but it does not become much smaller for about another third of the whole length, when the "small," as it is called, or tail, commences; and at this point also, on the back, is a large prominence of a pyramidal form, called the "hump," from which a series of smaller processes run half way down the "small," or tail, constituting what is called by whalers the "ridge." The body then contracts so much, as to become finally not thicker than the body of a man, and terminates by becoming expanded on the sides into the "flukes," or tail properly speaking. The two flukes constitute a large triangular fin, resembling in some respects the tail of fishes, but differing in being placed horizontally; there is a slight notch, or depression between the flukes, posteriorly–they are about six or eight feet in length, and from twelve to fourteen in breadth in the largest males. The chest and belly are narrower than the broadest part of the back, and taper off evenly and beautifully towards the tail, giving what by sailors is termed a "clear run,"–the depth of the head and body is in all parts except the tail greater than the width. The head viewed in front, as in fig.2, presents a broad, somewhat flattened surface, rounded, and contracted above, considerably expanded on the sides, and gradually contracted below, so as in some degree to attain a resemblance to the cutwater of a ship.

At the angle formed by the anterior and superior surfaces on the left side, is placed the single blowing-hole, or nostril, which in the dead animal presents the appearance of a slit or fissure, in form resembling as *S*, extending longitudinally, and about twelve inches in length.

This nostril, however, is surrounded by several muscles, which in the living state are for the purpose of modifying its shape and dimensions, according to the necessities of respiration, similar to those which act upon the nostrils of land animals.

In the right side of the nose, and upper surface of the head, is a large, almost triangular-shaped cavity, called by whalers the "case," which is lined with a beautiful glistening membrane, and covered by a thick layer of muscular fibres and small tendons, running in various directions, and finally united by common integuments. This cavity is for the purpose of secreting and containing an oily fluid, which, after death, concretes into a granulated substance of a yellowish colour, the spermaceti. The size of the case may be estimated, when it is stated that in a large whale it not unfrequently contains a ton, or more than ten large barrels of spermaceti!

Beneath the case and nostril, and projecting beyond the lower jaw, is a thick mass of elastic substance called the "junk:" it is formed of a dense cellular tissue, strengthened by numerous strong tendinous fibres, and infiltrated with very fine sperm oil and spermaceti.

The mouth extends nearly the whole length of the head. Both the jaws, but especially the lower, are in front contracted to a very narrow point, and when the mouth is closed, the lower jaw is received within a sort of cartilaginous lip, or projection of the upper one; but principally in front, for further back, at the sides, and towards the angle of the mouth, both jaws are furnished with tolerably well developed lips: in the lower jaw are forty-two teeth, of a formidable size, but conical shape; there are none, however, in the upper, which instead presents depressions corresponding to, and for the reception of, the points of those in the lower jaw,–sometimes, however, a few rudimentary teeth may be found situated in the upper jaw, but never projecting beyond the gums, and upon which those in the lower jaw strike when the mouth is closed.

The tongue is small, of a white colour, and does not appear to possess the power of very extended motion.

The throat is capacious enough to give passage to the body of a man; in this respect presenting a strong contrast with the contracted gullet of the Greenland whale.

The mouth is lined throughout with a pearly white membrane, which becomes continuous at the lips, and borders with the common integument, where it becomes of a dark-brown or black colour.

The eyes are small, in comparison with the size of the animal, and are furnished with eyelids, the lower of which is the more moveable: they are placed a little above, and behind the angle of the mouth, at the widest part of the head. At a short distance behind the eyes, are the external openings of the ears, of size sufficient to admit a small quill, and unprovided with any external auricular appendage.

Behind, and not far from the posterior angle of the mouth, are placed the swimming paws, or fins, which are analogous in their formation to the anterior extremities of other animals, or the arms of man; they are not much used as instruments of progression, but probably in giving a direction to that motion in balancing the body in sinking suddenly, and occasionally in supporting their young.

In a full-grown male sperm whale, of the largest size, or about eighty-four feet in length, the dimensions may be given as follow:–depth of head from eight to nine feet,–breadth, from five to six feet,–depth of body seldom exceeds twelve or fourteen feet, so that the circumference of the largest sperm whale of eighty or eighty-four feet will seldom exceed thirty-six feet,–the swimming paws or fins, are about six feet long and three broad; the dimensions of the flukes or tail have been previously mentioned.

In reviewing this description of the external form, and some of the organs of the sperm whale, it will perhaps not be uninteresting if some comparison is instituted between them and the corresponding points of the Greenland whale. In doing this, the remarkable adaptation of form and parts to different habits, situation, and food, will not fail to strike every one with admiration.

One of the peculiarities of the sperm whale, which strikes at first sight every beholder, is the apparently disproportionate and unwieldly bulk of the head; but this peculiarity, instead of being, as might be supposed, an impediment to the freedom of the animal's motion in his native element, is in fact, on the contrary in some respects very conducive to his lightness and agility, if such a term can with propriety be applied to such an enormous creature; for a great part of the bulk of the head is made up of a large thin membranous case, containing, during life, a thin oil of much less specific gravity than water;

below which again is the junk, which, although heavier than the spermaceti, is still lighter than the element in which the whale moves; consequently the head taken as a whole, is lighter specifically than any other part of the body and will always have a tendency to rise at least so far above the surface as to elevate the nostril or "blow-hole" sufficiently for all purposes of respiration, and more than this, a very slight effort on the part of the fish would only be necessary to raise the whole of the anterior flat surface of the nose out of the water; in case the animal should wish to increase his speed to the utmost, the narrow inferior surface, which has been before stated to bear some resemblance to the cutwater of a ship, and which would in fact answer the same purpose to the whale, would be the only part exposed to the pressure of the water in front, enabling him thus to pass with the greatest celerity and ease through the boundless track of his wide domain.

It is in this shape of the head that the sperm whale differs in the most remarkable degree from the Greenland whale, the shape of whose head more resembles that of the porpoise, and in it the nostril is situated much farther back, rendering it seldom or ever necessary for the nose to be elevated above the surface of the water; and when swimming even at the greatest speed, the Greenland whale keeps nearly the whole of the head under it, but as his head tapers off evenly in front, this circumstance does not much impede his motion, the rate of which is, however, never equal to that of the greatest rate of the sperm whale.

It seems, indeed, in point of fact, that this purpose of rendering the head of light specific gravity, is the only use of this mass of oil and spermaceti, although some have supposed, and not without some degree of probability, that the "junk" especially may be serviceable in obviating the injurious effects of concussion, should the whale happen to meet with any obstacle when in full career. This supposition, however, would appear hardly tenable, when we consider the Greenland whale, although living among the rock-like icebergs of the arctic seas, has no such convenient provision, and with senses probably in all, and certainly in one respect less acute that those of the sperm whale, on which account it would seem requisite for him to possess this defence rather than the sperm whale, whose habitation is for the most part in the smiling latitudes of the southern seas. Considering the habits and mode of feeding, and the superior activity and apparent intelligence of the sperm whale, we shall be prepared to expect that he must possess a corresponding superiority in his external senses; and we accordingly find, that he enjoys a more perfect organ of hearing, in having an external opening of considerable size for the purpose of conveying sounds to the internal ear more readily and acutely than could be done through the dense and thick integument, which is continued over the auricular opening in the northern whale.

Although the eyes in both animals are very small in comparison with their bulk, yet it is remarked that they are tolerably quick-sighted. I am not aware that the sperm whale possesses in this respect any superiority.

Passing to the mouth, we again observe a very remarkable difference in the conformation of the two animals; as in place of the enormous plates of whale-bone which are found attached to the upper jaw of the Greenland whale, we in the sperm whale only find depressions for the reception of the teeth of the lower jaw; organs which again are totally wanting in the other. Corresponding with these distinctions, which plainly point out that the food of the two whales must be very different, we find a remarkable difference in the size of the gullet.

The several humps, or ridges, on the back of the sperm whale constitute another difference in their external aspect; these prominences are however not altogether peculiar to the sperm whale, as that which is called by whalers the "humpback" possesses a prominence on the back not very dissimilar to that of the sperm whale, which has been noticed before in the introductory remarks, and which induced Lacapede to divide the genus Balaena into those with a hump, and those without; employing the name Balaena for the latter, and styling the others Balaenoptera.

I have before adverted to the sharp cutwater-like conformation of the under part of the head in the sperm whale, and it is worthy of remark that the same part of the Greenland whale is nearly, if not altogether, flat.

The skin of the sperm whale, as of all other cetaceous animals, is without scales, smooth, but occasionally, in old whales, wrinkled, and frequently marked on the sides by linear impressions, appearing as if rubbed against some angular body. The colour of the skin, over the greatest part of its extent, is very dark, most so on the upper part of the head, the back, and on the flukes, in which situation it is in fact sometimes black, on the sides it gradually assumes a lighter tint, till on the breast it becomes silvery grey.

In different individuals there is, however, considerable variety of shade, and some are even piebald. Old "bulls," as full-grown males are called by whalers, have generally a portion of grey on the nose immediately above the fore-part of the upper-jaw, and they are then said to be "grey-headed."

In young whales the "black skin," as it is called, is about three-eighths of an inch thick, but in old ones it is not more than one-eighth.

Immediately beneath the black-skin is the blubber or fat, which is contained in a cellular membrane, and which is much strengthened by numerous interlacements of ligamentous fibres, which has induced Professor Jacob to consider the whole thickness of blubber to be the *cutis vera,* or true skin, infiltrated with oil, or fatty matters. Its thickness on the breast of a large whale is about fourteen inches, and on most other parts of the body it measures from eight to eleven inches. The head is not, however, supplied with this covering, having only the black skin, or cutis, which lies close to a layer of very dense cellular tissue, under which is seen a considerable thickness of numerous small tendons, intermixed with muscular fibres.

This is more especially observed on the top and upper third of the head, surrounding the case, as lower down we find the black skin lying close to the peculiar structure of the junk.

This thick covering of skin, blubber, or fat, is called by the South Sea whalers the "blanket." It is of a light yellowish colour, and, when melted down, furnishes the sperm oil. It also serves two excellent purposes to the whale, in rendering it buoyant, and in furnishing it with a warm protection from the coldness of the surrounding element; in this last respect answering well to the name bestowed upon it by the sailors.

Ambergris

Although ambergris, even during the sixteenth century, appeared to be much valued as a mercantile commodity by the English, it is curious that we knew nothing of its source, and very little of the use which was made of it in other countries.

In the year 1672, we find the Hon. Robert Boyle claiming the honour of having discovered its source from a manuscript which was found on board a Dutch East Indiaman which had fallen into our hands by the chance of war. This precious document stated, that "ambergreese is not the scum or excrement of the whale, but issues out of the root of a tree, which tree, howsoever it stands on the land, alwaies [*sic*] shoots forth its roots towards the sea, seeking the warmth of it, thereby to deliver the fattest gum that comes out of it, which tree otherwise by its copious fatness might be burnt and destroyed: wherever that fat gum is shot into the sea, it is so tough that it is not easily broken from the root, unless its own weight and the working of the warm sea doth it, and so it floats on the sea; there was found by a souldier [*sic*] 7/8ths of a pound, and by the chief two pieces, weighing five pounds. If you plant the trees where the stream sets to the shore, then the stream will cast it up to great advantage! March 1st, 1672, in Batavia."–*Phil. Trans.,* vol. viii. p.6113.

But notwithstanding the above statement, Dr. Thomas Brown, in his work published a few years afterwards (1686), in his description of a sperm whale which was thrown on the coast of Norfolk, states that "in vain it was to rake for ambergriese in the paunch of this leviathan, as Greenland discoverers, and attests of experience dictate, that they sometimes swallow great lumps thereof in the sea–insufferable fetor denying that inquiry; and yet if, as Paracelsus encourageth, ordure makes the best musk, and from the most feted substances may be drawn the most odoriferous essences, all that had not Vespasian's nose might boldly swear here was a substance for such extractions;" which proves that the Dr. still suspected that the ambergris was found in the sperm whale although it was found by this animal floating in the sea, and swallowed by it in "great lumps!"

But it was reserved for Dr. Boylston, of Boston, to enlighten mankind on this important subject, and he therefore claims the discovery of its source in the following manner: "The most learned part of mankind are still at a loss about many things even in medical use, and particularly were so, in what is called ambergris, until our whale fishermen of Nantucket, in New England, some three or four years past made the discovery. Their account to me is this:–cutting up a spermaceti bull-whale, they found accidentally in him about twenty pounds' weight, more or less, of that drug; after which, they and other such fishermen became very curious in searching all such whales they killed, and it has been since found in lesser quantities in several male whales of that kind, and in no other, and that scarcely in one of a hundred of them. They add further, that it is contained in a syst [*sic*] or bag, without any inlet or outlet to it, and that they have sometimes found the bag empty and yet entire; the bag is nowhere to be found but near the *genital parts* of the fish. The ambergris is when first taken out moist, and of an exceedingly strong and offensive smell." This letter was written to the Royal Society in 1724.–*Phil. Trans.* vol.xxxiii. p.193.

In the same year, however, we have another letter from America, written to the Royal Society by the Hon. Paul Dudley, F.R.S., who, after telling us that the old sperm whales carry their young ones "on the flukes of their tails, who with their fins clasp about the small, and hold themselves on," also says "one of our country doctors tells me that the tooth of this fish (sperm whale) shaved or powdered, and so infused in liquor, equals the hartshorn, and has been used in the small-pox, and given to lying-in women in case of sickness, with success!–the quantity is a much as will lie upon an English shilling." Further on in the same letter he states, "I meddle not here with the *precious* ambergris found in this whale, because I design to close the whole with that discovery." And here is his conclusion: "But truth," says he, "is the daughter of time; it is now at length found out, that occultum naturae is an animal production, and bred in the body of the spermaceti whale. I doubt not," he continue, "but in process of time some further particulars may be procured with respect to ambergris, and I shall be proud to transmit them; in the mean time I hope the Society will accept of this first essay, and allow *my poor country* the honour of *discovering,* or at least ascertaining, the *origin* and *nature* of ambergris."–*Phil. Trans.* vol. xxxiii.

In a paper which was read before the Royal Society by Dr. Schwediawer, in 1783, respecting the medical properties of ambergris, he remarks, that "if we wish to see any medicinal effects from this substance, we must certainly not expect them from two or three grains, but give rather as many scruples of it for a dose; though even then I should not expect much from it, as I have taken of pure unadulterated ambergris in powder thirty grains at once, without observing the *least* sensible effect from it. A sailor, however, who had the curiosity to try the effects of some recent ambergris upon himself, took half an ounce of it melted upon the fire, and found it a good purgative, which proves that it is not quite inert."–*Phil. Trans.* vol. lxxiii. p.226.

In 1791, the attention of government was drawn to this subject, in order to discover if it could be more frequently found. When Captain Coffin was examined at the bar of the House of Commons on the subject, and stated that he had lately brought home 362 ounces, troy, of this costly substance, which he had found in the anus of a *female* sperm whale that he had captured off the coast of Guinea, and which he stated was very bony and sickly. At the time he brought this quantity to England, the ambergris was selling for 25*s.* an ounce, but he stated that he had sold his for 19*s.* 6*d.* per ounce, to a broker who exported it to Turkey, Germany, and France, among the natives of which it appears to have been long celebrated for its aphrodisiacal properties.

"The use of ambergris," says Brande, "in Europe is now nearly confined to perfumery, though it has formerly been used in medicine by many eminent physicians. In Asia and parts of Africa, ambergris is not only used as a medicine and perfume, but considerable use is also made of it in cooking, by adding it to several dishes as a spice. A great quantity of it is also constantly bought by the pilgrims who travel to Mecca, probably to offer it there, and make use of it in fumigations, in the same manner as frankincense is used in Catholic countries. The Turks make use of it as an aphrodisiac. Our perfumers add it to scented pastiles, candles, balls, bottles, gloves, and hair powder; and its essence is mixed with pomatum for the face and hands, either alone or united with musk, though its smell is to some persons extremely offensive. Ambergris may be known to be genuine by its fragrant scent when a hot needle or pin is thrust into it, and it melting like fat of a uniform consistence, whereas the counterfeit will not yield such a smell, nor prove of such a fat texture. One thing, however, is very remarkable, that a resemblance to the smell of this drug, which is the most agreeable of all the perfumes, should be produced by a preparation of one of the most odious of all substances. Mr. Homberg found that a vessel in which he had made a long digestion of human faeces, acquired a very strong and perfect smell of ambergris, inasmuch that any one would have thought that a great quantity of essence of ambergris had been made in it, the perfume was so strong and offensive that the vessel had to be removed from the laboratory!" *Brande's Manual of Chemistry,* p.594.

Ambergris appears to be nothing but the hardened faeces of the spermaceti whale, which is pretty well proved from its being mixed so intimately with the refuse of its food (the squids' beaks). Mr. Enderby has in his possession a fine specimen of this substance, about six or seven inches long, and which bears very evident marks of having been moulded by the lower portion of the rectum of the whale.

On one occasion, while in the North Pacific, I had the curiosity to collect some of the semi-fluid faeces which floated from the carcass of a whale while our men were cutting it in, and which on being dried in the sun bore all the properties of ambergris.

Chase and Capture of the Sperm Whale

Although ancient and modern histories may abound in descriptions of man's daring by "flood and field," and of the accidents and escapes which accompany his voluntary exposure to a multitude of dangers, surely the recital of his doings in the chase and capture of the leviathan of the deep–the great sperm whale–can be second to none in the interest it must excite in every contemplative mind. It is not in the field, jungle, or thick forest, that these hardy adventurers seed their prey, upon man's natural element,–and where, should any untoward accident occur, assistance of some kind can be readily obtained,–but on the vast ocean, at times thousands of miles distant from any habitable land, where he is not only exposed to the dangers which beset him in his adventures with the monster of the deep, but to others still more terrible, in which the ruthless tempest of the China and Japanese seas form no inconsiderable share; or when near lands, distant and barbarous, as far as the antipodes, dangerous reefs, sunken rocks, and relentless savages may beset him on every side, requiring all the moral and physical energy of which our nature is possessed, to escape the manifold dangers which surround them, but which the whale fisherman looks upon without dread, passing among them in his gallant bark, and carrying off in triumph the rich giant of the ocean.

When in pursuit of the whale with the boats, it occasionally happens that just at the moment the harpoon is about to be plunged into its body, the whale suddenly descends, leaving nothing but a vortex to mark the spot where but a moment before it was seen floating; but its course, however, has been observed, and the boats are placed in a position to be as near as possible to it when it again rises to breathe; the time, as has been before stated, when he will do this is known to a minute. If they should be more fortunate in the next rising of the whale, and they succeed in darting the harpoon into its body, then immediately after the first struggles of the wounded animal and when he is lying exhausted from his enormous exertions to escape, or free himself from the harpoon, the boat's head is placed close to its side, and the headsman begins to destroy it by thrusting his lance into its most vital parts, which lie near the fin, or darting at it from a distance; at the moment of lancing, he cries "stern all;" the oars are then immediately backed and the boat's stern becoming its cutwater, it is thus removed from danger without the loss of time and trouble in turning. Again, when feeling the lance, the whale plunges and throws itself in all directions, lashing the water with its tail, or rearing its enormous head, and threatening destruction with its formidable jaw. . . .

After being struck with the harpoon or lance, females and young bulls make the most violent efforts to escape, and being remarkably quick in their actions they frequently afford considerable danger and trouble. Those young bulls which yield about forty barrels of oil, and are consequently called forty-barrel bulls, are perhaps the most difficult to destroy, and sometimes make great havoc among the men and boats.

The large whales, such as make eighty or more barrels, not being nearly so active, and probably not feeling so acutely, are generally, by expert whalers, easily killed, and with less damage to those employed than the smaller ones. But these enormous creatures are sometimes known to turn upon their persecutors with unbounded fury, destroying everything that meets them in their course–sometimes by the powerful blows of their flukes, and sometimes attacking with the jaw and head. Accidents frequently occur from the violent convulsive movements of a wounded whale, when suffering the last pangs of his numerous and deep wounds. When the lance has been used effectively, so as to wound some important vital organ, the unfortunate animal frequently throws up blood in large quantities from the blow-hole–becomes convulsed–lashing the waves violently with its tail–passing very rapidly along–tinging the water with his blood as he swims; and it is a curious fact, that under these circumstances he always describes the segment of a circle: in this state the whale is said to be in his flurry, and some whalers state he always dies with his head towards the sun, but of this I have never been convinced. When dead, the body always turns on its side, probably from the greater weight of the spine and dorsal muscles.

In calm weather, great difficulty is sometimes experienced in approaching the whale, on account of the quickness of his sight and hearing–under these circumstances the fishers have recourse to paddles instead of oars, and by this means they with great caution sometimes get near enough to dart the harpoon with success. When first struck, the whale frequently "sounds," or descends to an amazing depth, taking out perhaps the lines belonging to the four boats, 800 fathoms! but afterwards, when weakened by loss of blood and fatigue, he becomes unable to sound to any great depth, and he then passes rapidly along the surface of the ocean, towing after him perhaps three or four boats. If he does not turn, the people in the boats draw in the line, by which they are attached to the whale, and thus easily come up with him even when going with great velocity, they then make use of the lance either by darting or thrusting, and so destroying their unoffending victim.

The scenes which sometimes occur during the chase and capture of this whale defy description. Let the reader suppose himself on the deck of a South-seaman, cruising in the North Pacific Ocean at its Japanese confine–he may be musing over some past event, the ship may be sailing gently along over the smooth ocean, every thing around solemnly still, with the sun pouring its intense rays with dazzling brightness; suddenly, the monotonous quietude is broken by an animated voice from the mast-head, exclaiming "there she spouts." The captain starts on deck in an instant, and inquires "where away!" but perhaps the next moment every one aloft and on deck can perceive an enormous whale lying about a quarter of a mile from the ship, on the surface of the sea, having just come up to breathe–his large "hump" projecting three feet out of the water, when at the end of every ten seconds the spout is seen rushing from the fore-part of his enormous head, followed by the cry of every one on board, who join heart and soul in the chorus of "there again!" keeping time with the duration of the spout. But while they have been looking, a few seconds have expired–they rush into the boats, which are directly lowered to receive them–and in two minutes from the time of first observing the whale, three or four boats are down, and are darting through the water with their utmost speed towards their intended victim, perhaps accompanied with a song from the headsman, who urges the quick and powerful plying of the oar, with the common whaling chant, of

"Away my boys, away my boys, 'tis time for us to go."

But while they are rushing along, the whale is breathing, they have yet perhaps some distance to pull before they can get a chance of striking him with the harpoon. His "spoutings are nearly out," he is about to descend, or he hears the boats approaching. The few people left on board, and who are anxiously watching the whale and the gradual approach of the boats, exclaim, "ah, he is going down!" yet he spouts again, but slowly, the water is again seen agitated around him, the spectators on board with breathless anxiety think they perceive his "small" rising in preparation for his descent; "he will be lost," they exclaim, for the boats are not yet near enough to strike him–and the men are still bending their oars in each boat with all their strength, to claim the honour of the first blow with their harpoon. The bow-boat has the advantage of being the nearest to the whale; the others, for fear of disturbing the unconscious monster, are now doomed to drop astern. One more spout is seen slowly curling forth,–it is his last, this rising,–his "small" is bent, his enormous tail is expected to appear every instant, but the boat shoots rapidly alongside of the gigantic creature. "Peak your oars," exclaims the mate, and directly they flourish in the air; the glistening harpoon is seen above the head of the harpooneer, in an instant it is darted with unerring force and aim, and is buried deeply in the side of the huge animal. It is "socket up;" that is, it is buried in his flesh up to the socket which admits the handle or "pole" of the harpoon. A cheer from those in the boats, and from the seamen on board, reverberates along the still deep at the same moment. The sea, which a moment before was unruffled, now becomes lashed into foam by the immense strength of the wounded whale, who with his vast tail strikes in all directions at his enemies. Now his enormous head rises high into the air, then his flukes are seen lashing everywhere, his huge body writes in violent contortions from the agony the "iron" has inflicted. The water all around him is a mass of foam, some of it darts to a considerable height–the sounds of the blows from his tail on the surface of the sea, can be heard for miles! . . .

"Stern all," cries the headsman; but the whale suddenly disappears; he has "sounded;" the line is running through the groove at the head of the boat, with lightning-like velocity; it smokes–it ignites, from the heat produced by the friction, but the headsman, cool and collected, pours water upon it as it passes. But an oar is now held up in their boat; it signifies that their line is rapidly running out; two hundred fathoms are nearly exhausted; up flies one of the other boats, and "bends on" another line, just in time to save that which was nearly lost. But still the monster descends; he is seeking to rid himself of his enemies by descending deeply into the dark and unknown depths of the fast ocean. They next bend on the "drougues," to retard his career,–but he does not turn; another and another have but slight influence in checking the force of his descent; two more lines are exhausted,–he is six hundred fathoms deep! "Stand ready to bend on," cries the mate to the fourth boat (for sometimes, though not often, they take the whole four lines away with them–800 fathoms!!); but it is not required, he is rising, "Haul in the slack," observes the headsman, while the boatsteerer coils it again carefully into the tubs as it is drawn up. The whale is now seen approaching the surface; the gurgling and bubbling water which rises before also proclaims that he is near; his nose starts from the sea; the rushing spout is projected high and suddenly, from his agitation. The "slack" of the line is now coiled in the tubs, and those in the "fast" boat haul themselves gently towards the whale; the boatsteerer places the headsman close to the fin of the trembling animal, who immediately buries his long lance in the vitals of the leviathan, while, at the same moment, those in one of the other boats dart another harpoon into his opposite side, when "stern all" is again vociferated, and the boats shoot rapidly away from the danger. . . .

Mad with the agony which he endures from these fresh attacks, the infuriated "sea beast" rolls over and over, and coils an amazing length of line around him; he rears his enormous head, and, with wide expanded jaw, snaps at everything around; he rushes at the boats with his head,–they are propelled before him with vast swiftness, and sometimes utterly destroyed.

He is lanced again, when his pain appears more than he can bear; he throws himself, in his agony, completely out of his element; the boats are violently jerked, by which one of the lines is snapped asunder; at the same time the other boat is upset, and its crew are swimming for their lives. The whale is now free! he passes along the surface with remarkable swiftness, "going head out;" but the two boats that have not yet "fastened," and are fresh and free, now give chase; the whale becomes exhausted, from the blood which flows from his deep and dangerous wounds, and the 200 fathoms of line belonging to the overturned boat, which he is dragging after him through the water, checks him in his course; his pursuers again overtake him, and another harpoon is darted and buried deeply in his flesh.

The men who were upset, now right their own boat without assistance from the others, by merely clinging on one side of her, by which she is turned over, while one of them gets inside and bales out the water rapidly with his hat, by which their boat is freed, and she is soon again seen in the chase.

The fatal lance is at length given,–the blood gushes from the nostril of the unfortunate animal in a thick black stream, which stains the clear blue water of the ocean to a considerable distance around the scene of the affray. In its struggles the blood from the nostril is frequently thrown upon the men in the boats, who glory in its show!

The immense creature may now again endeavour to "sound" to escape from his unrelenting pursuers; but it is powerless,–it soon rises to the surface, and passes slowly along until the death pang seizes it, when its appearance is awful in the extreme.

Suffering from suffocation, or some other stoppage of some important organ, the whole strength of its enormous frame is set in motion for a few seconds, when his convulsions throw him into a hundred different contortions of the most violent description, by which the sea is beaten into foam, and boats are sometimes crushed to atoms, with their crews. . . . But this violent action being soon over, the now unconscious animal passes rapidly along, describing in his rapid course a segment of a circle, this is his "flurry," which ends in his sudden dissolution. And the mighty rencontre is finished by the gigantic animal rolling over on its side, and floating an inanimate mass on the surface of the crystal deep,–a victim to the tyranny and selfishness, as well as a wonderful proof of the great power of the *mind* of man.

In the afternoon of a day which had been rather stormy, while we were fishing in the North Pacific, a "school" of young bull-whales made their appearance close to the ship, and as the weather had cleared up a little, the captain immediately ordered the mate to lower his boat, while he did the same with his own, in order to go in pursuit of them.

The two boats were instantly lowered, for we were unable to send more, having had two others "stove" the day before; they soon got near the whales, but were unfortunately seen by them before they could get near enough to dart the harpoon with any chance of success, and the consequence was that the "pod" of whales separated, and went off with great swiftness in different directions. One however, after making several turns, came at length right towards the captain's boat, which he observing, waited in silence for his approach without moving an oar, so that the "young bull" came close by the boat, and received the blow of the harpoon some distance behind his "hump," which I saw enter his flesh myself, as it occurred close to the ship. The whale appeared quite terror-struck for a few seconds, and then suddenly recovering itself, darted off like the wind, and spun the boat so quickly round when the tug came upon the line, that she was within a miracle of being upset. But away they went, "dead to windward," at the rate of twelve or fifteen miles an hour, right against a "head sea," which flew against and over the bows of the boat with uncommon force, so that she at times appeared ploughing through it, making a high bank of surf on each side.

The second mate, having observed the course of the whale and boat, managed to waylay them, and when they came near to him, which they speedily did, a "short warp" was thrown, and both boats were soon towed at nearly the same rate as the captain's boat had been before.

I now saw the captain darting the lance at the whale as it almost flew along, but he did not seem to do so with any kind of effect, as the speed of the whale did not appear in the least diminished, and in a very short time they all disappeared together, being at too great a distance to be seen with the naked eye from the deck. I now ran aloft, and with the aid of a telescope could just discern from the mast-head the three objects, like specks upon the surface of the ocean, at an alarming distance. I could just observe the two boats, with the whale's head occasionally darting out before them, with a good deal of "white water" or foam about them, which convinced me that the whale was still running. I watched them with the glass until I could no longer trace them, even in the most indistinct manner, and I then called to those on deck, that they might take the bearing by compass, of the direction in which I had lost sight of them, that we might continue to "beat" the ship up to that quarter.

It was now within half-an-hour of sunset, and there was every appearance of the coming on of an "ugly night," as a seaman would say; indeed the wind began to freshen every moment, and an "awkward bubble" of a sea soon began to make. I remained aloft until I saw the sun dip, angry and red, below the troubled horizon, and was just about to descend when I was dreadfully shocked at hearing the loud cry of "a man overboard" from all upon deck. I immediately looked astern, and saw one of our men, of the name of Berry, grappling with the waves and calling loudly for help. The ship was soon brought round; but in doing so she unavoidably passed a long way from the poor fellow, who still supported himself by beating the water with his hands, although he was quite unacquainted with the proper art of swimming. Several oars were thrown overboard the moment after he fell, but he could not reach them, though they were near to him; and directly the ship was brought up, a Sandwich islander, who formed one of the crew, leaped overboard and swam towards him, while at the same time the people on the deck were lowering a spare boat, which is always kept for such emergencies. I could be of no service, except to urge their expedition by my calls, for it was all only the work of a few minutes. The good Sandwich islander struck out most bravely at first; but in a short time, finding that he was some distance from the ship, and being unable to see Berry, on account of the agitated surface of the sea, actually turned back through fear–finding, as he said, that the "sea caps" went over his head. The men in the boat now plied their oars with all their strength, and were making rapidly towards the drowning young man, who now and then disappeared entirely from view under the heavy seas which were beginning to roll; a sickening anxiety pervaded me, as my thoughts appeared to press the boat onwards to the spot where the poor fellow still grappled, but convulsively, with the yielding waters. The boat, urged by man's utmost strength, sprang over the boisterous waves with considerable speed; but they arrived half a minute too late to save our poor shipmate from his watery grave. I saw him struggle with the waves until the last, when the foam of a broken sea roared over him, and caused him to disappear for ever! The boat was rowed round and round the fatal spot, again and again, until night fell, and then she was slowly and reluctantly pulled to the ship by her melancholy crew. As they returned, the turbulent waves tossed them about, as if in sport, dashing its spray at times completely over them, making the boat resound from the beating of the dashing waters which flew against her bow.

The moment the unfortunate seaman disappeared, a large bird of the albatros [*sic*] kind came careering along, and alighted on the water at the very spot in which the poor fellow was last seen. It was a curious circumstance, and only served to heighten our horror, when we saw this carnivorous bird seat itself proudly over the head of our companion, and which also served to remind us of the number of sharks that we had so frequently seen of late, of the horrible propensities of which we could not dare to think.

By the time we had hoisted in the boat it was quite dark; the winds too had increased to half a gale, with heavy squalls at times, so that we were obliged to double-reef our topsails. Our painful situation now bore most heavily upon us. We had lost one of our men, who had sailed with us from England–the bare thought of which in our circumstances aroused a crowd of heart-rending ideas. Our captain and second mate, with ten of the crew, had also disappeared, and were by this time all lost, or were likely to be so in the stormy night which had now set in; being too several hundred miles away from any land. We however, kept beating the ship to windward constantly, carrying all the sail that she could bear, making "short boards" or putting the ship about every twenty minutes. We had also, since nightfall, continued to burn blue lights, and we had likewise a large vessel containing oil and unravelled rope, burning over the sternrail of the ship as a beacon for them, which threw out a great light. But although all eyes were employed in every direction searching for the boats, no vestige of them could be seen; and therefore when half-past nine P.M. came, we made up our minds that they were all lost; and as the wind howled hoarsely through the rigging, and the waves beat savagely against our ship, some of us thought we could hear the shrieks of poor Berry above the roaring of the storm; others imagined in their melancholy, that they could occasionally hear the captain's voice, ordering the ship to "bear up," while the boats had been seen more than fifty times by anxious spirits, who had strained their eyes through the gloom until fancy robbed them of their true speculation and left her phantasmagoria in exchange. There were not many on board who did not think of home on that dreadful night–there were not many among us who did not curse the sea, and all sea-going avocations; while, with the same breath, they blessed the safe and cheerful fireside of their parents and friends who resided at home, and which at that moment they would have given all they possessed but to see. But at the moment despair was firmly settling upon us, a man from aloft called out that he could see a light right a head of the ship, just as we were "going about," by which we should have gone from it. We all looked in that direction, and in a few minutes we could plainly perceive it; in a short time we were close up with it, when, to our great joy, we found the captain and all the men in the boats, lying to leeward of the dead whale, which had in some measure saved them from the violence of the sea. They had only just been able to procure a light, having unfortunately upset all their tinder through the violent motion of the boats, by which it became wet–but which they succeeded in igniting after immense application of the flint and steel–or their lantern would have been suspended from an oar directly after sunset, which is the usual practice when boats are placed under such circumstances.

After having secured the whale alongside, and which we expected to lose during the night from the roughness of the weather, they all came on board, when the misfortune of poor Berry was spoken of with sorrow from all hands, while the joy of their own deliverance served to throw a ray of light amidst the gloom.

On the morning of the 18th June, 1832, while we were still fishing in the "off-shore ground" of Japan, we fell in with an immense sperm whale, which happened to be just the sort of one we required to complete our cargo. Three boats were immediately lowered to give him chase; but the whale, from some cause or other, appeared wild in its actions long before it had seen any of our boats, although it might have been chased the day before by some other ship. It was greatly different in its actions to most other large whales, because it never went steadily upon one course. If he "peaked his flukes," or went down going to the southward, we expected he would continue that course under water, but when he again rose perhaps he was two or three miles away from the boats to the northward; in this sort of manner he dodged us about until near four P.M., at which time the men were dreadfully exhausted from their exertions in the chase, which had been conducted under a broiling sun, with the thermometer standing in the shade at 93°. About half-past four, however, Captain Swain contrived, by the most subtle management and great physical exertions, to get near to the monster, when he immediately struck him with the harpoon with his own hands; and, before he had time to recover from the blow, he managed with his usual dexterity to give him two fatal wounds with the lance, which caused the blood to flow from the blow-hole in abundance. The whale, after the last lance, immediately descended below the surface, and the captain felt certain that he was going to "sound," but in this he was much mistaken–for a few minutes after his descent he again rose to the surface with great velocity, and striking the boat with the front part of his head threw it high into the air with the men and everything contained therein, fracturing it to atoms and scattering its crew widely about. While the men were endeavouring to save themselves from drowning by clinging to their oars and pieces of the wreck of the boat, the enormous animal was seen swimming round and round them, appearing as if meditating an attack with his flukes, which if he had thought proper to do, in return for the grievous wounds that he had himself received, a few strokes of his ponderous tail would soon have destroyed his enemies; but this was not attempted. They had now nothing to hope for but the arrival of the other boats to relieve them from their dangerous situation, rendered more so by the appearance of several large sharks, attracted by the blood which flowed from the whale, which were sometimes only a few feet from them; and also from the inability of one of the boats' crew to swim, by which three or four of his mates were much exhausted in their efforts to save him, which they succeeded in doing after having lashed two or three oars across the stern of the boat, which happened to be not much fractured, on which they placed their helpless fellow-adventurer. After they had remained in the water about three quarters of an hour, assisting themselves by clinging to pieces of the wreck, one of the other boats arrived and took them in, no doubt greatly to their relief and satisfaction. But although these brave whale fishermen had been so defeated, they were not subdued: the moment they entered the boat which took them from the ocean, their immediate determination was for another attack upon the immense creature, which remained close by, while the other boat, which was pulling towards them with all the strength of its rowers, would still be a quarter of an hour before it could arrive.

Captain Swain, with twelve men in one boat, therefore made another attack upon the whale with the lance which caused it to throw up blood from the blow-hole in increased quantities. We, who were on board the ship, and had observed from a great distance, by means of the telescope, the whole of the occurrence, were employed in beating the ship towards them, but they were far to windward, and the wind being rather light, we had even our royal sails set. Soon after the arrival of the third boat, the whale went into its flurry and soon died, when, to the dismay of the boats' crews, who had endured so much danger and hardship in its capture, it sunk, and never rose again,–an occurrence which is not very unfrequent, owing of course to the greater specific gravity of the individual, perhaps from a greater development of bony and muscular structures. Such were the adventures of that day, in the evening of which the crews returned to the ship, worn out and dispirited, having lost a favourite boat, with the whole of her instruments, besides the last whale wanted to complete the cargo, and worth at least 500£! . . .

At day-break, one fine morning in August, as our first mate was going aloft, to look out for whales, he discovered no less than three ships within a mile of us; but they were situated in various directions. We soon discovered them to be whalers, who like ourselves were cruising after the spermaceti whale, and therefore their appearance only had the effect of redoubling our vigilance in the "look-out," so that we might if possible be the first, to obtain the best chance if one of those creatures "hove in sight." And it was not long before a very large whale made his appearance right in among the ships. The water was smooth at the time, for we had but a light air of wind stirring, so that our boats were instantly lowered without the loss of time of bringing the "ship to." But although we managed matters as quietly and secretly as possible, we found the moment our boats cleared the ship's side, that all the others had been as vigilant as ourselves, and had also lowered their boats after the whale. The whole of them immediately began the chase,–nine boats in all, being three from each ship. They all exerted themselves to their utmost, and as we expected, in vain; for before any of the boats had got even near him, the enormous animal lifted his widely expanded flukes, and descended perpendicularly into the depths of the ocean to feed. Those in the boats, however, having noticed his course, proceeded onwards, thinking the whale would continue to pursue the same direction under water; but as he was going slowly at the time he was up, they did not proceed more than a mile from the place at which he descended before they separated about a hundred yards from each other, and then "peaking" their oars, all the men in each boat stood up, looking in different directions, so as to catch the first appearance of the spout when the whale again rose to breathe: when an hour after his descent had expired, the excitement among us who were on board the ship, became wound up to its highest pitch. The captain, who had remained on board, ascended to the fore-top-gallant-yard to watch the manoeuvres of the boats, and for the purpose of the better ordering the signals to them, or working of the ship. All those who were down after the whale, appeared as feverish with anxiety as ourselves, for every now and then they were to be seen shifting their positions a little, thinking to do so with advantage; then they would cease rowing, and stand upon the seats of the boats, and look all round over the smooth surface of the ocean with ardent gaze. But one hour and ten minutes expired before the monster of the deep thought proper to break cover, and when he did, then a rattling chase commenced with the whole of the boats, and they really flew along in fine style, some of them actually appearing to be lifted quite on the surface of the water from the great power of the rowers; and we had the satisfaction of observing, that our boats were quite equal to the others in the speed with which they were propelled. But it was again a useless task, as the old "schoolmaster" had outwitted those in the boats, by having gone–while under water–much further than any of his pursuers had anticipated, and they again had the mortification of witnessing the turning of his flukes as he once more descended into the depths of his vast domain. We now knew to a minute the time that he would remain below, while the people in the boats continued to row slowly onwards the whole time. A fine breeze now sprang up, so that we were enabled to keep company with the boats, keeping a little to windward of them, as the whale was going "on a wind," as a seaman would say, meaning that it was blowing across him. When the hour and ten minutes had again nearly past, the nine boats were nearly abreast of each other, and not much separated, so that the success of first striking the whale depended very much upon the swiftest boat, especially if the whale came up a-head. We had now all the boats on our "lee beam," while the ships were all astern of us, and the most distant not being more than half a mile, so that we enjoyed an excellent view of this most exciting and animated scene. True to his time, the leviathan at length arose right a-head of the boats, and at not more than a quarter of a mile distant from them. The excitement among the crews of the various boats when they saw his first spout was tremendous; they did not shout, but we could hear an agitated murmur from their united voices reverberating along the surface of the deep. They flew over the limpid waves at a rapid rate; the mates of the various boats cheered their respective crews by various urgent exclamations, "Swing on your oars my boys, for the honour of the Henrietta," cried one; "Spring away, hearties," shouted another, and yet scarcely able to breathe from anxiety and exertion. "It's our fish," vociferated a third, as he passed the rest of his opponents but a trifling distance. "Lay on my boys," cried young Clark, our first mate, as he steered the boat with one hand, and pressed down the after oar with the other. "She'll be ours yet; let's have a strong pull, a long pull, and a pull altogether," he exclaimed, as he panted from his exertions at the after oar, which soon brought up his boat quite abreast of the foremost. But the giant of the ocean, who was only a short distance before them, now appeared rather "gallied," or frightened, having probably seen or heard the boats, and as he puffed up his spout to a great height, and reared his enormous head, he increased his speed, and went along quite as fast as the boats,–but for only two or three minutes, when he appeared to get perfectly quiet again, while the boats gained rapidly upon him, and were soon close in his "wake." "Stand up," cried young Clark to the harpooneer, who is also the bow-oarsman, while the same order was instantly given by his opponent, whose boat was abreast of our mate's, with the rest close to their sterns. The orders were instantly obeyed, for in a second of time both boatsteerers stood in the bows of their respective boats, with their harpoons held above their heads ready for the dart; but they both panted to be a few yards nearer to the whale, to do so with success. The monster ploughed through the main quickly, but the boats gained upon him every moment, when the agitation of all parties became intense, and a general cry of "dart! dart!" broke from the hindermost boats, who each urged their friends, fearful of delay. The uproar became excessive, and while the tumult of voices, and the working and splashing of the oars, rolled along the surface of the deep, both the harpooneers darted their weapons together, which, if they had both struck the whale, would have originated a contention between them regarding their claims, But, as it happened, neither of them had that good fortune; for at the moment of their darting the harpoons, the whale descended like a shot, and avoided their infliction, leaving nothing but a white-and-green-looking vortex in the disturbed blue ocean, to mark the spot where his monstrous form so lately floated. A general huzza burst from the sternmost boats when they saw the issue of this chase, thinking now that another chance awaited them on the next rising of the whale, and they soon began to separate themselves a little, and to row onwards again in the course which they thought he had taken. Our captain, feeling irritated at the ill-success of the mate, now ordered his own boat to be lowered, intending to make one in the chase himself; but, just as he had parted from the ship, going down a little to leeward, a tremendous shout arose from their people in our own boats, joined with a loud murmuring from the rest of the boats' crews; for the whale, not having had all its "spoutings out," had now risen again to finish them, and was coming to windward at a quick rate, right towards our ship. The captain saw his favourable situation in a moment, and passing quickly to the bows of the boat, he stood to waylay him as he came careering along, throwing his enormous head completely out of the water, for he was not quite "gallied." He soon came, and caught a sight of the boat just as he got within dart; the vast animal rolled itself over in an agony of fear, to alter its course; but it was too late, the harpoon was hurled with excellent aim, and was plunged deeply into his side near the fin. As the immense creature almost flew out of the water from the blow, throwing tons of spray high into the air, shewing that he was "fast," a triumphant cheering arose from those in our own boats, as well as from those in the ship, accompanied by exclamations loud and deep, and not of the most favourable kind to us, from all the rest. But onwards they all came, and soon cheerfully rendered assistance to complete its destruction; but which was not done, however, without considerable difficulty, the whale continuing to descend the moment either of the boats got nearly within dart of him. But after an hour's exertion in this way, six out of the ten boats which were now engaged got fast to him by their harpoons, but not one of them could get near enough to give him a fatal lance; he towed them all in various directions for some time, taking care to descend below the surface the moment a boat drew up over his flukes, or otherwise drew near, which rendered it almost impossible to strike him in the body, even when the lance was darted, although the after part of his "small" was perforated in a hundred places: from these wounds the blood gushed in considerable quantities, and as the poor animal moved along, towing the boats, he left a long ensanguined stain in the ocean. At last, becoming weak from his numerous and deep wounds, he became less capable of avoiding his foes, which gave an opportunity for one of them to pierce him to the life!–dreadful was that moment the acute pain which the leviathan experienced, and which roused the dormant energies of his gigantic frame. As the life's blood gurgled thick through the nostril, the immense creature went into his "flurry" with excessive fury, the boats were speedily sterned off, while he beat the water in his dying convulsions with a force that appeared to shake the firm foundation of the ocean!

Numberless stories are told of fighting whales, many of which however, are probably much exaggerated accounts of the real occurrences. A large whale, called "Timor Jack" is the hero of many strange stories, such as of his destroying every boat which was sent out against him, until a contrivance was made by lashing a barrel to the end of the harpoon with which he was struck, and whilst his attention was directed and divided amongst several boats, means were found of giving him his death wound.

In the year 1804, the ship "Adonis," being in company with several others, struck a large whale off the coast of New Zealand, which "stove" or destroyed nine boats before breakfast, and the chase consequently was necessarily giving up. After destroying boats belonging to many ships, this whale was at last captured, and many harpoons of the various ships that had from time to time been sent out against him were found sticking in his body. This whale was called "New Zealand Tom," and the tradition is carefully preserved by whalers.

Accidents of the most fearful nature have frequently occurred in this hazardous pursuit, which to enumerate would fill the space of volumes;–for not only boats, but sometimes even ships have been destroyed by these powerful creatures. It is a well authenticated fact, that an American whale-ship called the "Essex" was destroyed in the South Pacific Ocean by an enormous sperm whale. While the greater part of the crew were away in the boats, pursuing whales, the few people remaining on board saw an immense sperm whale come up close to the ship, and when very near he appeared to go down for the purpose of avoiding the vessel, and in doing so he struck his body against some part of the keel, which was broken off by the force of the blow and floated to the surface; the whale was then observed to rise a short distance from the ship, and come with apparently great fury towards it, striking against one of the bows with his head, and completely "staving" it in. The ship of course immediately filled, and fell over on her side, in which dreadful position the poor fellows in the boats soon espied their only home, being distant from the nearest land many hundred miles; on returning to the wreck they found the few who had been left on board hastily congregated in a remaining whale boat, into which they had scarcely time to take refuge before the vessel capsized. They with much difficulty obtained a scanty supply of provisions from the wreck, their only support on the long and dreary passage before them to the coast of Peru, to which they endeavored to make the best of their way. One boat was fortunately found by a vessel not far from the coast; in it were the only survivors of the unfortunate crew, three in number, the remainder having perished under unheard-of suffering and privation. These three men were in a state of stupefaction, allowing their boat to drift about where the winds and waves listed; one of the survivors was the master; by kind and careful attention on the part of their deliverers, they were eventually rescued from the jaws of death to relate the melancholy tale.

From James Colnett, *A Voyage to the South Atlantic and round Cape Horn into the Pacific Ocean, for the Purpose of Extending the Spermaceti Whale Fisheries*

(London: W. Bennett, 1798)

INTRODUCTION

James Colnett (1753–1806) had sailed as a midshipman with Captain James Cook (1728–1799) on the latter's second voyage of discovery to the Pacific Ocean (1772–1775) and devoted most of his subsequent career to the exploration of the west coast of North America.

For a book about extending the whale fishery, Colnett's account is slim on whaling. It is rich in other detail, however, and looms large in *The Encantadas*, either directly or via Porter's *Journal*. Colnett's drawing of a sperm whale is often reproduced, but seldom with the text. Oddities in the text of the plate are likely due to the process of engraving, which is relatively unforgiving of mistakes in spelling, capitalization, or punctuation. Since Colnett appears conspicuously in Melville's other early whaling sources, it is highly likely that he began to search for a copy as soon as he borrowed Scoresby and ordered Beale—in late April 1850. Bercaw #154. Modern edition: Da Capo, 1968.

PRIMARY SOURCE

Narborough Isle is the highest land among the Galipagoe [*sic*] Islands, lying near the center of Albemarle Isle, which almost surrounds it, in the form of two crescents, and making two bays. The apparent point of division of these islands, is so low on both, that I am in doubt whether they are seperated [*sic*]. On the next morning we saw spermaceti whales, we killed seven and got them along side [*sic*]; Rock Rodondo bearing East 22° South, the Northernmost land bearing East 18° East. The weather was hazy, and the Latitude by observation 00° 27' 13" North. Here we cruised till the eighth of April, and saw spermaceti whales great numbers, but only killed five, of which we secured four. The current ran so strong to the Westward, and the winds were so light, that after laying to, to secure the whales and cut them up, we were seven days in returning to the ground from whence we drifted. In the winter season, when the winds are more fresh, these difficulties might not occur, otherwise, it would be impossible for any vessel, which was not a very prime sailer, to whale here with success; though at a certain season any quantity of sperm oil might be procured. The oldest whale-fishers, with whom I have conversed, as well as those on board my ship, uniformly declared that they had never seen spermaceti whales in a state of copulation, or squid their principal food in shoals before; but both these objects were very common off these isles, and we frequently killed the latter, of four or five feet in length, with the granes. Young spermaceti whales were also seen in great numbers, which were not larger than a small porpoise. I am disposed to believe that we were now at the general rendezvous of the spermaceti whales from the coasts of Mexico, Peru, and the Gulf of Panama, who come here to calve: as among those we killed, there was but one bull-whale. The situation I recommend to all cruizers [*sic*], is between the South end of Narborough Isle and the Rock Rodondo: though great care must be taken, not to go to the North of the latter; for there the current sets at the rate of four and five miles an hour due North. Narborough Isle falls gradually down to a point at the North, South, and East ends, and may be equal in produce to any of the neighbouring isles; but of this I can only conjecture, as I did not myself examine it; nor does it appear that the Buccaneers ever landed upon it.

The Rodondo is an high barren rock, about a quarter of a mile in circumference, and is visible as far as eight or nine leagues, has soundings round it at the distance of a quarter of a mile thirty fathom. Here our boats caught rock-cod in great abundance. I have frequently observed the whales leave these isles and go to the Westward, and in a few days, return with augmented numbers. I have also seen the whales coming, as it were, from the main, and passing along from the dawn of day to reach the Galipagoes. It is very much to be regretted that these isles have to this period been so little known but only to the Spaniards. . . .

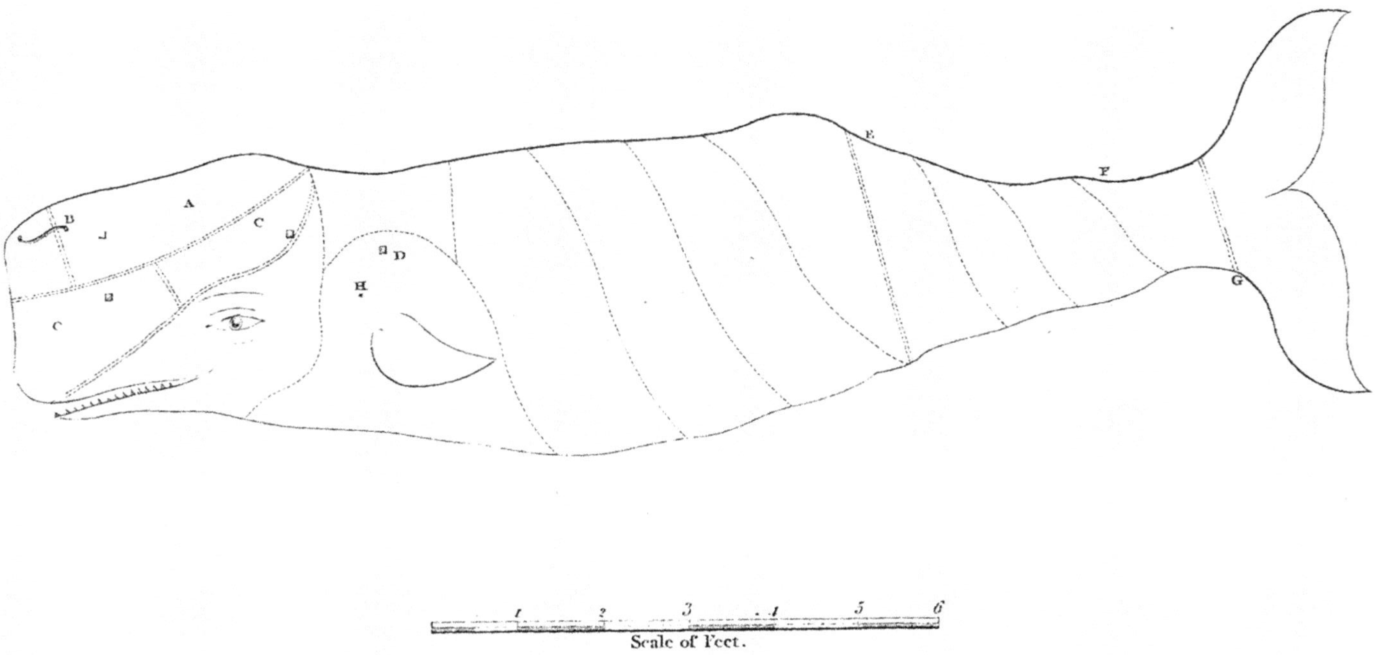

"Physeter, or Spemaceti Whale," from Colnett (1798) (James Colnett, *A Voyage to the South Atlantic and round Cape Horn into the Pacific Ocean, for the Purpose of Extending the Spermaceti Whale Fisheries* (London: W. Bennett, 1798))

A. Part of the Head containing liquid Oil which is covered with a black membrane.

B. The Spout hole which runs horizontally along the left side, and is also seperated by the same kind of membrane. The part between the two double lines is covered with Fat of considerable thickness, like that of a hog, these parts make one third of the quantity of Oil the Fish produces, of which the liquid is about one third.

A.B. Part of the Head which of large Whales being too bulky and ponderous to be hoisted on board is suspended in tackles and the front part cut off as described thus, and the Oil bailed out with buckets; but in small Whales the head is divided at the double line below CC. and hoisted upon deck.

■■ Where the tackles are toggled or hook'd.

D. Where the tackles are first hooked, which is called raising a piece, being thus steadied in the tackles the head is divided at the lowest double line and were a stern till the fish is flinched, which is done by seperating the Fat from the Body with long handled Iron Spades, as the Whale is hove round by the tackles the Fat peels off, and if any Sea is on the rising of the Ship considerably expedites the business.

E. A large lump of Fat.

F. A smaller. When the Fish is flinched, or peeled to E. it will no longer cant in the tackles, is therefore cut through at the first double line and also at G. the Tail being of no value.

H. The Ear, which is remarkably small in proportion to the body as is also the Eye from which a hollow or concave line runs to the fore part of the head the Eyes being prominent enables them to pursue their Prey in a direct line, and by inclining the head a little either to the right or left to see their enemy a stern, they have only one row of Teeth, which are in the lower Jaw with sockets in the upper one to receive them, the number depends on the age of the Fish, the lower Jaw is a solid bone that narrows nearly to a point and closes under the upper, when they spout they throw the water forwards and not upwards like other Whales except when they are enraged, they also spout more regular and stay longer under water the larger the Fish the more frequently they spout and continue longer under water. The Tail is horizontal with which he does much mischief in defending himself. Their Food, from all the observations I have had an opportunity of making, has been the Sepia or middle Cuttle Fish. This species of the Whale, is remarkable for its attachment and for assisting each other when struck with a harpoon: and more mischief is done by the loose Fish, than those the boats are fast to, and they frequently bite the lines in two which the struck Fish is fast with. The Ambergrease is generally discovered by probing the intestines with a long Pole, when the Fish is cut at E.

From Horatio Hastings Weld, *Ribs and Trucks, from Davy's Locker*

(Boston: Charles D. Strong, 1842)

INTRODUCTION

Shortly after publishing his book under the initials "W.A.G." Weld (1811–1888) became an Episcopalian minister. Well might Weld have published under a pseudonym: the book is a collection of light journalistic pieces of the least substantial kind, of which the first is a specious account of whaling as an exercise in yachting. In the 1840s, Weld also edited *Brother Jonathan*, a periodical read in the Melville household. Bercaw #752b.

PRIMARY SOURCE

'Tis your whaler alone, who goes down *to* the sea, in ships; other mariners hurry across it. He alone does business *upon* the great waters; and, more emphatically, than other "sea-farers," makes the ocean his home. With his top-sail-yard "sharp up," and his helm "four spokes a-lee," he rides out the storm, month in and month out—nestling, like a sea bird, in the trough, and feeling himself, as it were, in harbor; while other ships rise on the horizon and scud by, and lessen and reel out of sight, with blind celerity, on their respective courses. A trip to Europe may serve to introduce the novice to the "blue deity;" but, for a thorough acquaintance, commend me to the long, familiar intercourse of a whaling voyage.

From William Dampier, *A New Voyage round the World*

(London: James and John Knapton, 1729)

INTRODUCTION

The real Dampier (1651–1715) is best known as being the cousin of the fictional Lemuel Gulliver in Jonathan Swift's (1667–1745) *Travels into Several Remote Nations of the World,* better known as *Gulliver's Travels* (1726). It is still uncertain whether Melville knew the works of the buccaneer Dampier at firsthand or only through his reading of other collections of voyages. Dampier was energetically interested in natural history, and yet this particular voyage is remarkable for its *lack* of observations of whaling. An essential work if one were writing about, say, tortoise hunting, Dampier's book mentions whales only through two brief discussions of ambergris. And yet, as the excerpts below suggest, the book may well foreshadow "Chapter 110: Queequeg and his Coffin." Bercaw #188. Modern Edition: New York: Da Capo, 1970.

PRIMARY SOURCE

Upon this Occasion, I cannot omit to tell my Reader what I learnt from Mr. *Hill* the Chirurgeon, upon his shewing me once a piece of Ambergreece [*sic*], which was thus. One Mr. *Benjamin Barker*, a Man that I have been long well acquainted with, and know him to be a very diligent and observing Person, and likewise very sober and credible, told this Mr. *Hill,* that being in the Bay of *Honduras* to procure Logwood, which grows there in great abundance, and passing in a Canoa over to one of the Islands in that Bay, he found upon the Shoar, on a sandy Bay there, a lump of Ambergreece, so large, that when carried to *Jamaica*, he found it to weigh a Hundred Pound and upwards. When he first found it, it lay dry, above the Mark which the Sea then came to at High-water; and he observed in it a great Multitude of Beetles: It was of a dusky colour, towards black, and about the hardness of mellow Cheese, and of a very fragrant Smell: This that Mr. *Hill* shewed me, being some of it which Mr. *Barker* gave him. Besides those already mentioned, all the Places where I have heard that Ambergreece [*sic*] hath been found, at *Bermudas*, and the *Bahama* Islands in the *West-Indies*, and that part of the Coast of *Africk* with its adjacent Islands, which reaches from *Mozambique* to the *Red-Sea*. . . .

After Four a-Clock the Thunder and the Rain abated, and then we saw a *Corpus Sant* at our Main-top-mast Head, on the very Top of the Truck of the Spindle. This sight rejoiced our Men exceedingly; for the height of the Storm is commonly over when the *Corpus Sant* is seen aloft; but when they are seen lying on the Deck, it is generally accounted a bad Sign.

A *Corpus Sant* is a certain small glittering Light; when it appears as this did, on the very Top of the Main-mast or at a Yard-arm, it is like a Star; but when it appears on the Deck, it resembles a great Glow-worm. The *Spaniards* have another Name for it, though I take even this to be a *Spanish* or *Portuguese* Name, and a Corruption only of *Corpus Santum*) and I have been told that when they see them, they presently go to Prayers, and bless themselves for the happy Sight. I have heard some ignorant Seamen discoursing how they have seen them creep, or, as they say, travel about in the Scuppers, telling many dismal Stories that hapned [*sic*] at such times: but I did never see any one stir out of the place where it was first fixt [*sic*], except upon Deck, where every Sea washeth it about: Neither did I ever see any but when we have had hard Rain as well as Wind; and therefore do believe it is some Jelly: but enough of this. . . .

Ambergreece is often found by the Native *Indians* of these Islands, who know it very well; as also know how to cheat ignorant Strangers with a certain Mixture like it. Several of our Men bought such of them for a small Purchase. Captain *Weldon* also about this time touched at some of these Islands, to the North of the Island where we lay; and I saw a great deal of such Ambergreece, that one of his Men bought there; but it was not good, having no smell at all. Yet I saw some there very good and fragrant. . . .

Thus it was that I came to have this painted Prince, whose Name was *Jeoly*, and his Mother . . . When I showed him Spice, he would not only tell me that there was *Madochala*, that is, abundance; but to make it appear more plain, he would also show me the Hair of his Head, a thing frequent among all the

Indians that I have met with, to show their Hair, when they would express more than they can number. He told me also, that his Father was *Raja* of the Island where they lived: That there were not above Thirty Men on the Island, and about one Hundred Women: That he himself had five Wives and eight Children, and that one of his Wives painted him.

He was painted all down the Breast, between his Shoulders behind; on his Thighs (mostly) before; and in the Form of several broad Rings, or Bracelets round his Arms and Legs. I cannot like the Drawings to any Figure of Animals, or the like; but they were very curious, full of great variety of Lines, Flourishes, Chequered-Work, &c. keeping a very graceful Proportion, and appearing very artificial, even to Wonder, especially that upon and between his Shoulder-blades. By the Account he gave me of the manner of doing it, I understood that the Painting was done in the same manner, as the *Jerusalem* Cross is made in Mens Arms, by pricking the Skin, and rubbing in a Pigment. But whereas Powder is used in making the *Jerusalem* Cross, they at *Meangis* use the Gum of a Tree beaten to Powder, called by the *English,* Dammer, which is used instead of Pitch in many parts of *India.* He told me, that most of the Men and Women on the island were thus painted. . . .

I had no Employment for them; but they both employed themselves. She used to make and mend their own Cloaths [*sic*], at which she was not very expert, for they wear no Cloaths at *Meangis,* but only a Cloath about their Wastes: And he busied himself in making a Chest with four Boards, and a few Nails that he begged of me. It was but an ill-shaped odd Thing, yet he was as proud of it as if it had been the rarest Piece in the World.

From Joseph C. Hart, *Miriam Coffin, or the Whale-Fishermen*

(New York: Carvill, 1834)

INTRODUCTION

Hart (1798–1855) is best known for being the first critic to suggest that William Shakespeare did not write the plays that were published under his name. Melville found Hart's *The Romance of Yachting* (1848) so appalling that he refused to review it, but scholars since the 1920s have emphasized Melville's borrowings from the earlier work in *Moby-Dick*. From the nature of his comments on *The Romance of Yachting,* it seems unlikely that Melville was aware of *Miriam Coffin* in the autumn of 1848. Hart's whaling novel was dedicated to Boston Loyalist Sir Isaac Coffin, Baronet (1759–1839), who endowed a school on Nantucket. Bercaw #325. Modern edition: New York: Garret Press, 1969.

PRIMARY SOURCE

During a tour to the eastward, several years ago, among the clusters of small islands which lie off the southern shore of Massachusetts, and which have proved the great nurseries of seamen devoted to the whale-fishery, it was our good fortune to sojourn for a season upon Nantucket. The principal object of our journey, at that time, being to obtain authentic information in relation to the actual state of that important trade, we embraced every opportunity of conversing with those who pretended to knowledge respecting it; and, in our necessary intercourse with all classes, it frequently happened that common sailors, who had spent their lives in whaling vessels, furnished the best sources of information—especially as we found many owners and captains who regarded us with shyness, whenever we broached the subject of the mystery of the trade "whereby they did live" and had procured their gains. Among other things, we were anxious to discover by what means it had been brought to the present perfection in its management, and had maintained its prosperity under the absolute neglect of the General Government, whose province and whose constitutional duty it is to give facility to all our branches of commercial trade and navigation.

We were not surprised, however, at the discovery, upon our arrival at Nantucket, that its people had occasionally suffered from this neglect. We found that they had already presented a memorial to Congress, setting forth the great hazards, and complaining of the unprotected situation of their trade, which they had for half a century prosecuted in the Pacific Ocean,—where their ships were not only subject to dangers in navigation, from the uncertainty of the surveys and the inaccuracy with which many islands had been laid down on the charts, but were constantly liable to vexatious detention and exactions from the authorities upon the South American coast. The petitioners expressed their desire that Government would take the matter in hand, and fit out a small naval expedition, which should be directed to cruise in the neighbourhood of their whaling-grounds; and otherwise be employed occasionally in making accurate surveys of new places, that were already or might thereafter be discovered; as well as to ascertain the capabilities of the islands for affording those natural and necessary supplies which all whale-ships are in want of in the course of their long voyages among numberless isles, where nature has been all-bountiful in the over-abundance of her productions, but where the tenant of the soil is, in the same proportion, rude and inhospitable.

The short memorial of the Nantucket people, whose prosperity at home is so closely linked with the success of the whale-fishery in remote seas, represents:—"That the intercourse maintained between different ports of the nation and the islands and countries of the Pacific Ocean has become a matter of public interest, and deserving the protecting care of the National Legislature. The fur business, and the trade carried on between the Pacific islands and coasts of China, have afforded rich returns, and increased the wealth of our common country. Besides this employment of national industry and enterprise, they would represent, that there are engaged in the whale-fishery, from various parts of the country, upwards of forty thousand tons of shipping, requiring a capital of three millions of dollars, and the services of more than three thousand seamen. Whether viewed as a nursery of bold and hardy seamen, or as an employment of capital in one of the most productive modes, or as furnishing an article of indispensable necessity to human comfort, it seems to your petitioners to be an object especially deserving the public care. The increased extent of the voyages now pursued by the trading and whaling ships, into seas but little explored, and to parts of the world before unknown, has increased the cares, the dangers, and the losses of our merchants and mariners. Within a few years, their cruises have extended from the coasts of Peru and Chili to the North-west Coast, New-Zealand, and the isles of Japan. This increase of risk has been attended by an increase of loss. Several vessels have been wrecked on islands and reefs not laid down on any chart; and the matter acquires a painful interest from the fact that many ships have gone into those seas, and no soul has survived to tell their fate. Your petitioners consider it a matter of earnest importance that those seas should be explored; that they should be surveyed in an accurate and authentic manner, and the position of new islands, and reefs, and shoals definitely ascertained. The advancement of science, and not their private interest only, but the general interest of the nation, seem to them imperiously to demand it.

They therefore pray that an expedition may be fitted out, under the sanction of the Government, to explore and survey the islands and coasts of the Pacific seas." . . .

In a secluded quarter of the island of Nantucket, known by the name of Siasconset, there lived, a few years since, a singular being, whose mode of life, for several previous years, had been a mystery to everybody. To this individual, however, we had been directed for information on a point embraced in our investigations, respecting the state of the whale-fishery as connected with Nantucket. He had been represented by the people of the town as possessing a remarkably retentive memory,—particularly in what related to the early history of the island; and also that he was possessed of large stores of accurate statistical and historical information, which he had been many years in collecting and arranging; and furthermore it was reported, that in his person one might discover a walking genealogical tree, whose leaves and branches, so to speak, would unfold the birth, parentage, and education of every resident of the island, from the days of the first settlers downwards to the time present.

There are now some three or fourscore houses at Siasconset, of one story and a half in height, erected on the margin of a high sand-bluff overlooking the sea. Some of these are very old, and built after a peculiar fashion which prevailed all over the island during the early part of the last century. It was then a small village, inhabited by poor fishermen, and the huts we speak of were their domicils. Latterly, however, these huts have been turned into summer residences for the wealthier townspeople; and right pleasant lounging places do they make, for those who have leisure to enjoy them. If any of our readers should feel curious to see the style of building that prevailed one hundred years ago in the town which has since assumed the name of Nantucket, let him now pay a visit to Siasconset, and enter its dwellings, and regard attentively its pepper-box outhouses. He will there see how of old every inch of room was economized, and how sleeping-chambers were scaled by perpendicular step-ladders, like those used to descend to the pent-up cabin of a fishing-smack, or to clamber up the sides of a merchantman; and how the best and most spacious room in the house is finished like the cabin of a ship, with projecting beams, whose corners are beaded and ornamented with rude carving, while the walls are wainscoted with unpainted panel-work, and the oaken floors have grown alike brown by time, and smooth by a century's use. There is but one house in the whole village which makes modern pretension to fashionable exterior. It is the only innovation upon the unity—the ancient "*keeping*" of the place; and its projector deserves banishment under the wise provisions of the time-honoured "*Laws of 'Sconset*" for presuming to make any change in the architecture of the settlement.

It was our fortune to make a pilgrimage to Siasconset at that season of the year when its houses were tenantless,—its deserted avenues choked up with sombre and lifeless thistles and decayed long grass,—and all as still as the grave. Threading with uncertainty its narrow and silent lanes, in search of the habitation of the veteran, we came at length to a hut before whose door stood a car of fish, which had been recently caught and wheeled up from the shore. The chimney-top, too, gave evidence of civilization and of the whereabout of humanity. A stream of blue smoke issued forth, and briskly curled up in the clear atmosphere. The sight of the fish, jumping and floundering about in the little car, and the lively jet of smoke overhead, was as welcome to us, at the moment, as a house of "entertainment for man and beast" would be to a traveller in the desert, or to a virtuoso, without corn in his scrip, exploring the mysteries and antiquities of a city of the dead. We tapped lightly on the closed door of the hut, and repeated the signal more than once: but no answer from the indweller bade us welcome to the hospitalities of 'Sconset.

"This is strange!" thought we,—"very strange in a land proverbially celebrated for the open door and the open hand!"

A thirst after knowledge, and a stomach yearning fearfully for a morsel from the frying-pan or the fish-pot, gave us the courage of desperation; and thereupon we lifted the latch of the door,—for lock, or bolt, or other fastening, there was none,—and entered boldly into the main apartment of the house. There we stood for some minutes, silently contemplating the furniture and appointments of the place. It was clear that the hand of woman had not been there for many a day, though it was evident, from the arrangement of pots, and kettles, and platters, and frying-pans, that attempts had been made, if not with female neatness, at any rate with manly clumsiness and good-will, to preserve a degree of cleanliness that was creditable to the owner of the mansion. Over the rude mantel hung an old-fashioned turnip-shaped silver watch, ticking loudly, and striving on in its daily race with the sun; and against the still ruder partition, which separated the larger room from a closet or sleeping apartment, hung a heavy fowling-piece of most capacious bore; while underneath depended a well worn shot-bag, and a powder-flask of semi-transparent form. Around the room, somewhat in confusion, the implements of piscatory warfare were visible. Scap-nets and fishing-lines, of various sizes and lengths, wet from recent use, were spread over the backs of chairs to dry, and indicated that their owner had but lately come from an excursion upon the sea.

There was no help for us but to sit down and quietly await the approach of the master, and the issue of our adventure. On coming to this very natural conclusion, we drew the only chair which was disengaged towards the ingulfing fireplace, and essayed to correct the chilled atmosphere of the room, by feeding the decaying fire with billets from a small heap of prepared wood piled in the corner, which, from certain appearances, had been gathered along the beach, and had once formed a part of some unfortunate vessel wrecked upon the shoals of the island.

There we sat, punching the fire with the tongs, and watching the sparks "prone to fly upwards," and wondering where all this would end. A dreamy sort of abstraction came over our faculties; and in this secluded spot we almost began to fancy that we were alone in the world. We felt some of those sensations creeping upon us which one might suppose the last man would feel, who had seen all generations pass into the grave,—leaving him the sole tenant of the earth. The crooked legs and claw-feet of the little old-fashioned cherry table multiplied a thousand fold in number and in crookedness, till we almost fancied it a huge creeping thing, with the legs and arms and claws of a dragon.

Presently an agonized groan escaped from the chest of some sufferer near at hand, and invaded the deep silence of the place,—which before had been rendered doubly painful by the distant monotonous roar of the surf, rolling and tumbling in upon the beach. We dropped the tongs in affright; and mechanically springing upon our feet, we were in the act of rushing forth from the cabin, to avoid the perturbed ghost which our imagination had conjured up to haunt the place withal.

"Who's there?" said a loud voice that appeared to come from the cockloft.

The charm was at once broken by the utterance of these words in the vernacular tongue, and our nervous sensations gave way before the idea of the utter ridiculousness of running away under such circumstances. We had always longed for solitude,—for "a lodge in some vast wilderness,"—but that charm, too, was broken; and we believed, in our very souls, that we had had enough of the eternal silence, which is too often hankered after by the "mind diseased."

"Henceforth," said we, mentally, "give us the hum and the bustle of the world, and the sprightly chat of intimacy. Solitude!—thus do we blow thee to the winds!"

We answered the hail from aloft, nothing loath; and begged the host to come down, as we had walked full seven miles to see and converse with him upon matters with which he was reputed to be familiar. The burley form of the man now darkened the aperture above, and he descended the step-ladder, with his back towards us, holding on for safety, and letting himself down with both hands, by two knotted cords,—such as are thrown over at the gangway of a man-of-war, to aid the descent into the tiny cutter alongside. As he stood confronting us, we could not fail to observe that he must have seen many winters and some hardships. His face was much weather-beaten, and his head, bald in some spots, was here and there covered with long and thin tufts of whity-grayish locks, standing up and streaming out in admirable confusion. Deep boots, resembling fire-buckets, together with drab small-clothes, encased his legs; while his upper garments were covered over with a huge shaggy wrapper, which sailors call a monkey-jacket. He looked at us keenly for a moment; but finding his craft fairly boarded, and in possession of the enemy, he deigned to offer us a seat, and to utter an excuse for his absence, by telling us that he had sought rest in his chamber after the fatigues of his late excursion. Moreover, he explained the cause of his fearful groaning, by giving a graphic portrait of the fiend-like nightmare which the falling of the tongs had scared away from his breast. We did not, upon the whole, find our companion as morose as we had been led to believe, by the description given to us of his habits. At any rate, he gradually became familiar, and undertook to find out for us, Heaven knows by what intricate process, a collateral descent from the "*great Trustum Coffin*;" and perhaps to this circumstance, more than to any other, are we indebted for the favours, both of speech and manuscript, which he afterward bountifully showered upon us.

"Odds-fish!" exclaimed he of the monkey-jacket, breaking in upon a long historical descent, in the mazes of which he had involved himself while answering a casual question of ours—"odds-fish! thou must have fasted sufficiently well by this late hour; and I will defer giving the remainder of the information which thou hast demanded, until our frugal meal is prepared and discussed. I have but few luxuries, friend—what didst call thy name?"

"Thompson, sir," said we at a venture, feeling for the present a desire to preserve our incognito.

"Thompson, is it?—I thought thou saidst but now it was Jenkins."

"Thompson, sir—a relative of the Jenkinses by the mother's side."

"Ah—well—I have but few luxuries, friend Thompson, to offer thee in this mine humble abode; but if, peradventure, thou art fond of fish, and bringest a good appetite, I will prepare thee such a dish as the townspeople can scarcely make without a resort to 'Sconset." Whereupon our companion selected a large fish from his car, and in a trice disrobed it of its scales and disembowelled the intestines; while, in order to gain some little credit for skill in culinary handiwork, and furthermore to convince him that we knew how to accommodate ourself to circumstances (or that, in the words of a Jonathan in the east, "while in Turkey we could do as the Turkeys did," we seized upon a bucket, and filled it with the purest of water at the village- pump;—and then we kindled up the fire anew, and made all things ready for the accommodation of the dinner-pot.

In due time, but not a minute too soon, a savoury dish of chowder came upon the table; and such is the force of a good appetite, we did think that in all our life before we had never swallowed provender half so delicious. But let that pass: the reader whose mouth waters must go to 'Sconset for his chowder, if he would, like unto us, enjoy a superlative luxury compounded of simples.

As the clam-shell dipper, which had come and gone full oft between our pewter platters and the chowder-pan, rested from its labours, the host pushed back his chair. Whereupon, lighting his pipe, and coming to an anchor in his easy-chair in the corner, he cast his eyes up towards the well-smoked roof in a sort of thinking revery, and at last broke silence as follows:—

"As I was telling thee, friend Tompkins, the island that now bears the name of Nantucket, whose barren plains thou hast crossed in coming hither, was once a well-wooded and well-watered garden-spot. It was owing to the improvidence, or perhaps I might better say to the lack of foresight, of our ancestors, that every tree of native growth, save one or two little clumps of oak, hath disappeared from the face of our land. It is melancholy to think on't—for I love the sight of trees. The soil, however, friend Tompkins, as thou mayst have observed, is not altogether as steril as the world in general imagine. But the cry of the 'sand-heap' hath gone out against us:—and herein I would say something to thee about evil speaking; but of that hereafter, if we have time.

"To make a long story short, friend Timpson," continued the narrator, "I will give thee merely the outline of our history, which, as time and opportunity serve, thou mayst fill up at leisure. Nay—do not interrupt me—I will answer thee more at large upon any point thou mayst propose when my sketch is finished. Being a stranger here, it may profit thee to know, that for a long time after the cession of the colony of New-York to Lord Stirling, the island of Nantucket, as well as all other islands on the northern coast, were claimed as dependencies of that distant colony. It came to pass, however, that by peaceable negotiation Massachusetts obtained dominion over the islands upon her shore, and Block Island fell to the lot of the Providence Plantations; while Long Island, with which nature had defended the shore of Connecticut, continued the appendage of New-York.

"Touching the manner in which Nantucket was settled by the whites, I have authority for declaring that it was brought about by accident, as it were, and under peculiar circumstances. We, who are natives of the island, trace our descent to the Seceders, or rather to the Non-Conformists, who dwelt in the eastern part of Massachusetts. They were principally of the Baptist persuasion; and, in ancient times, they were persecuted and hunted down by their Puritanic brethren, for opinion's sake. By one of those strange inconsistencies incident to human nature, the Puritans upon the main, who had themselves been the objects of persecution in England, began the same infamous and brutal career of intolerance in America, by establishing a code of revolting laws, which would have put a Herod to the blush. I thank God, my friend, that I am not descended from that vile fanatical race. Let others boast, if they will, of their Puritanic blood,—mine knows not the contamination!"

Here my companion rose from his chair, and opened a tobacco-closet in the chimney-side, from whence he produced a well-thumbed volume, and read as follows:—

"'No Quaker, or dissenter from the worship of the established dominion, shall be allowed to give a vote for the election of magistrates, or any officer.

"'No food or lodging shall be afforded a Quaker, Adamite, or other heretic.

"'If any person turns Quaker, he shall be banished, and not suffered to return but on pain of death.

"'No Roman Catholic priest shall abide in the dominion; he shall be banished, and suffer death on his return.'

"Such, my friend," continued our host, "were the laws of the Cameronians; and to their existence may be attributed the settlement of Nantucket, as thou wilt presently see. About the year 1650–60, while these and other fiend-like enactments were in force in the eastern section of the present United States, one Thomas Macy, a Baptist, who had come from England some twenty years previous, in search after a peaceful habitation in our Western wilds, and who had settled among the Puritans at Salisbury in the Massachusetts, committed a crying sin against the laws of the wrathful Cromwellites, or Blueskins. And what think'st thou it was? He had dared to shelter some forlorn and houseless Quakers in his barn one tempestuous night; and for that offence was he doomed by the Puritanic Roundheads to undergo the signal punishment of stripes at the whipping-post! Before the day of its infliction arrived, he procured an open boat, or yawl, and with two companions, Edward Starbuck and a youth by the name of Isaac Coleman, he launched forth upon an unknown sea,—declaring that he would pull his bark to the ends of the earth, sooner than dwell longer among beings so uncharitable and intolerant.

"Macy and his friends arrived at Nantucket, where before the white man had never dwelt. At that time two hostile tribes of Indians inhabited opposite ends of the island, numbering altogether some three thousand souls. The new comers were received with kindness by the natives; and they obtained a great but honest influence over their councils. Thus commenced the settlement of Nantucket by the whites; and in the following year one Thomas Mayhew, having obtained a grant of the island from Lord Stirling, conveyed it, in fee, to ten proprietors, each of whom chose an associate from among his brother 'heretics;' and the whole company of twenty, with their persecuted families, immediately thereafter took possession as proprietors in common."

Our companion hereupon pulled forth a slip of paper from a long-worn pocket-book, from which we took the liberty of transcribing the names of the original settlers of the island. Although some of the names are now extinct, we would preserve the remainder, if possible, to their posterity. Their industry, single-mindedness, and perseverance are worthy of the admiration and the imitation of their descendants.

The First Ten

Thomas Mayhew,
Thomas Macy,
Tristram Coffin,
Thomas Barnard,
Peter Coffin (son of Tristram),
Christian Hussey,
Stephen Greenleaf,
John Swain,
William Pile,
Richard Swain.

Their Associates

John Smith,
Edward Starbuck,

Nath'l Starbuck (son of Edw'd),
Robert Barnard,
James Coffin (brother of Peter),
Robert Pike,
Tristram Coffin, jr.,
Thomas Coleman,
Nathaniel Bolton,
Thomas Losk.

Finishing the transcript of these venerable names, we handed back to our companion the original list. He took the paper between his finger and thumb, and with his nail resting on the third name from the top, he remarked, with a glow of pride, that the direct descendants of the senior Tristram Coffin had been computed at the enormous number of twenty-five thousand!—A prolific progenitor, and a goodly posterity, truly.

We now ventured to start a theme upon which our host dilated with wonderful fluency and apparent delight; and, in the course of a short time, we were made acquainted with the history of the rise and progress of the whale-fishery in these parts. But as the reader may, in other places in these pages, find the subject touched upon by an abler pen than ours, and perchance derive an interest from the perusal great as our own, we will omit the detail here; merely premising, however, that the daring natives of the island of Nantucket, in their frail canoes, first initiated the white settlers in the dangerous art of grappling with

"That sea-beast
Leviathan, which God of all his works
Created hugest that swim the ocean-stream."

But the ingenious whites, who in the beginning dared the perils of the sea in their little open skiffs, and dashed in among the huge game that then played about the island in troops, soon betook themselves to larger vessels, and made lengthened voyages upon the ocean, out of sight of the island. From such imperfect and perilous beginnings have they come to be the most hardy and expert whalers in the world.

Feeling anxious to ascertain by what means the first settlers had been converted from the faith of the Baptists to that of the Quakers, we gave a hint to that effect, and were enlightened accordingly.

"Quakerism," said he, "was not introduced upon the island until about the year 1701; when John Richardson, an itinerant but powerful and accomplished

Quaker preacher, came among the people. Mary Starbuck, eldest daughter of the 'first Trustum Coffin,' was not only the first English child born upon the island, but the first convert to the Quaker faith. The Other settlers gradually embraced the peaceful doctrines of Fox and Barclay, as preached by Richardson, and eventually the Society of Friends became the predominant sect. To the introduction of Quakerism, and its unvarying customs, together with the unyielding manners of the Puritans to which the islanders had been accustomed, and which still lingered about them even after their change of faith, added to their isolated situation, is perhaps to be attributed the unchangeableness of the ways and habits of the Nantucket people. It is the last hold of the simple manners of our English ancestors in America. In this respect, Nantucket is to the rest of America what Iceland is to the northern nations of Europe. For while all has undergone a change, I will not say for the better, in the continental countries, these two islands continue to exhibit the manners, customs, dress, and language of their ancestors, in much of their pristine purity.

"The spirit of resistless change is, however, abroad in Nantucket" (here the narrator heaved a sigh, and continued); "and I grieve to say the few years last past have worked a wonderful change in the people. The Indian prophecy hath come to pass in a shape which our fathers little dreamed of. Thou must know, my friend, that when the pestilence raged among the natives of the island in the year 1764, which reduced their numbers to a mere handful, but left the whites unscathed, the noble blue-fish, such as thou hast this day partaken of, disappeared entirely from our waters. It is now more than threescore years since the species was thought to have become extinct. The superstitious natives looked upon the unaccountable disappearance of the blue-fish, which previously they had caught in immense numbers, as the sure forerunner of the total extinction of their Indian race;—and it was even so. 'But,' said they, in bitterness, 'when our fire is extinguished, and our wigwams have become razed, then the blue-fish will return. Then let the shad-belly and the long-tail, (as they called the Quakers) look out for his dwelling and his landmarks, and that the stranger wrest not his inheritance from him as he has wrested ours from us!'

"Now mark me, my friend," solemnly and slowly continued our companion, "mark what I tell thee in relation to the Indian prophecy:—the blue-fish *have* returned within the present year,—the *last* Indian lingers among us, without the hope of issue,—and the places of the wigwams of his fathers are only known by their desolated hearths. The lineal descendants of the original proprietors are scattered over the world, and are disappearing from among us before the face of strangers who have come into the isle. Our broad corn-fields are trodden down, and our 'Shearing' scarce deserves the name;—not a single custom of our ancestors is adhered to in its ancient purity,—all, all is giving way before the spirit of innovation that now stalks abroad in the island,—prostrating all that is venerable for its antiquity, and good as being the delight of our fathers!"

An honest tear came to the eye of the old man as he closed his historical details.

We confess that the ready information furnished by our host had made the time pass away with unwonted celerity. With reluctance we cast our eyes out upon the sun, which was fast running down the west; and we reached for our hat, and held out our hand to take leave. Our companion gave it a kindly pressure, and followed us to the door with our hand folded in his.

"Stop!" exclaimed he, suddenly; "do not go yet. Take it not as flattery, but I am pleased with thy curiosity and thy intelligence. I would fain bestow a mark of my favour upon thee; and the more readily and willingly, because thy conversation hath been both amusing and instructive. I would crave, therefore, a repetition of thy visit to this lone dwelling, from which the idle and the impertinent have been excluded for years."

While our kind host was once more engaged in ransacking his tobacco-closet, we endeavoured to recall the portions of our conversation in which we had conveyed the least information in the world, or in which we had rendered ourself at all amusing; but, certes, we could not recollect saying more than ten words at any one time,—and those were put in edgewise, in the shape of questions; and truly do we believe that our host uttered ninety-and-nine full sentences to every monosyllable of ours. The thought flashed upon us that we had been an attentive listener—and the secret was out! Men given to be garrulous always praise good listeners.

Great was our surprise when our new-made friend approached and put into our hands a ponderous roll of papers, carefully tied up with a piece of tarred rope-yarn.

"There!" said our host of the monkey-jacket; "take it, friend Tinker" (this was the fourth time he had miscalled the name)—"take it, friend Tinker, and mend it if thou wilt. Peradventure some pestilent printer, like him at the town, may use his types upon it, instead of printing essays upon schools and temperance, as *he* hath done ('ad rat him!), to make children wiser and better than their fathers.[1] There is truth in every page of that manuscript, my friend; and moreover something about the perils of the whale-fishery which I have been a matter of twenty winters in putting together, after an experience and observation of more than sixty years:—and I have hoped the while, that it might some day be instrumental in bringing back to the people of my native island the recollection of the golden days of their ancient customs, from the which, alas! they have greatly departed of late, to cleave unto the fashions and vanities of the great cities. I have shaken the dust from my feet, in testimony against their multiplied follies; and have come out from among them, more in sorrow than in anger, to dwell here alone upon the sea-shore. But fare thee well, friend—how dost call thy name again?"

"Thompson, sir."

"Ah—Thompson!—I shall remember it when thou comest again. Once more, fare thee well!"

We now turned our back upon the little village, and made our way with rapid strides towards the town. The sun was sinking in the ocean as we commenced our retrograde march over the heath, and the full moon danced upon the waters ere we regained our hotel. In our eagerness to inspect the package so singularly committed to our charge, we thought not upon the weariness attendant upon a seven miles' tramp; and putting aside, rather unceremoniously, the cup of refreshing souchong tendered to us by the kind mistress of the mansion, we seized a lamp from the mantel and hastened to our chamber. We cut the matted rope-yarn which secured the bundle with an unskilfulness that deprived one of "Rogers's Best" of its keen edge, which half a day's friction upon the "Franklin Hone," and a faithful strapping upon the "Remedy for Wry Faces" to boot, with difficulty restored.

The severed string unfolded to our eyes the titlepage of the following Tale; and upon the next leaf we discovered the Dedication which the reader has found prefixed to this present writing. . . .

Note

1. We presume the allusion here made was to Mr. Jenks, the spirited editor of the Nantucket Inquirer, whose faithful labours in the cause of education should be esteemed above all praise. His zeal may have given offence to some of the old-school gentlemen, who, like our narrator, were opposed to innovation of every sort, and were content with the "humanities'" as taught by such ancient dames as the "Widow Cradders," and "Mary Gardner," and "Nabby Bunker,"—names well remembered at Nantucket,—who, it is related, suffered their pupils to go to sleep comfortably throughout the hours of their school sessions. But they have been long gathered to their fathers. We marvelled why, in denouncing the "pestilent printer," he did not also give a thrust at Admiral Coffin, to whom he had dedicated his work, and who had, previous to the above interview, established the foundation of the "*Coffin Grammar School*" at Nantucket, with a most munificent endowment out of his own

private funds. It is matter of wonder why he did not address the worthy admiral in the words of *Cade*:—

"Thou hast most, traitorously corrupted the youth of the realm in erecting a grammar school. It will be proved to thy face that thou hast men about thee that usually talk of a noun and a verb, and such abominable words, as no Christian ear can endure to hear."

Posterity, however, will do justice to the motives of Mr. Jenks and Admiral Coffin: and we mistake if the present island generation do not regard their public labours with a proper appreciation.

Editor.

From Obed Macy, *The History of Nantucket*

(Boston: Hilliard, Gray, 1835)

INTRODUCTION

Obed Macy (1762–1844) was no less a historian of whaling than he was of the island of Nantucket. His title is misleading: his account of the whale fishery is in no way a subordinate part of the book, but is almost the *raison d'etre* of it. And yet Macy's book is more than occasionally left out of bibliographies of the most important whaling books because of the opacity of the short title. The full title of Macy's book is *The History of Nantucket; being a compendious account of the first settlement of the island by the English, together with the rise and progress of the whale fishery; and other historical facts relative to said island and its inhabitants.* A "Note to the Reader" seems to claim partial authorship, or at least a division of editorial tasks, for William Coffin (?1756–1835).

In this account, some things are abbreviated, some perhaps even "stupedly" [*sic*] so, as Melville complained about popular accounts of the *Essex* tragedy. There is no mention of cannibalism beyond the words "indulgence" and "imagination." The shooting of a man in Captain Pollard's boat is omitted. Even the names of the dead are erased: this account is no Seaman's Bethel memorial or six-inch chapter to the dead (and eaten). Wonderfullest things are ever the unmentionable, as Melville wrote in "The Lee Shore" in *Moby-Dick*. But to be fair, Macy was essentially writing for the friends and relatives of these same dead and eaten victims: one can understand his reticence.

Melville's surviving copy is inscribed and dated 7 1/m [January] 1852; references within *Moby-Dick* are from either a secondary source or an earlier copy. Sealts #345; Bercaw #469.

PRIMARY SOURCE

Loss of the Ship Essex

A very interesting narrative of the loss of this ship, and the consequent sufferings of the crew, was published a few years since, which we recommend to the perusal of our readers. In the following account we shall confine ourselves to only a few of the events consequent to the singular accident, which occasioned the loss of the Essex.

The ship Essex, Captain George Pollard, sailed from Nantucket, 12th of 8th month, 1819, on a whaling voyage to the Pacific Ocean. Her crew consisted of 21 men, 14 of whom were whites, mostly belonging to Nantucket, the remainder were blacks. On the 20th of the 11th month, 1820, in lat. 0° 40' S. lon. 119° W. a school of whales was discovered, and in pursuing them the mate's boat was stove, which obliged him to return to the ship, when they commenced repairing the damage. The captain and 2d mate were left with their boats pursuing the whales. During this interval the mate discovered a large spermaceti whale near the ship, but, not suspecting the approach of any danger, it gave them no alarm, until they saw the whale coming with full speed towards them. In a moment they were astonished by a tremendous crash. The whale had struck the ship a little forward of the fore chains. It was some minutes before the crew could recover from their astonishment, so far as to examine, whether any damage had been sustained. They then tried their pumps, and found that the ship was sinking. A signal was immediately set for the boats. The whale now appeared again making for the ship, and, coming with great velocity, with the water foaming around him, he struck the ship a second blow, which nearly stove in her bows. There was now no hope of saving the ship, and the only course to be pursued was, to prepare to leave her with all possible haste. They collected a few things, hove them into the boat and shoved off. The ship immediately fell upon one side and sunk to the water's edge. When the captain's and 2d mate's boats arrived, such was the consternation, that for some time not a word was spoken. The danger of their situation at length aroused them, as from a terrific dream, to a no less terrific reality. They remained by the wreck two or three days, in which time they cut away the masts, which caused her to right a little. Holes were then cut in the deck, by which means they obtained about 600 pounds of bread, and as much water as they could take, besides other articles likely to be of use to them. On the 22d of 11th month, they left the ship, with as gloomy a prospect before them as can well be imagined. The nearest land was about 1,000 miles to the windward of them; they were in open boats, weak and leaky, with a very small pittance of bread and water for the support of so many men, during the time they must necessarily be at sea. Sails had been prepared for the boats, before leaving the ship, which proved of material benefit. They steered southerly by the wind, hoping to fall in with some ship, but in this they were disappointed. After being in their boats 28 days, experiencing many sufferings by gales of wind, want of water, and scanty provisions, they arrived at Ducie's Island, in lat. 24° 40' S. lon. 124° 40' W. where they were disappointed in not finding a sufficiency of any kind of food for so large a company to subsist on. Their boats being very weak and leaky, they were hauled on shore and repaired. They found a gentle spring of fresh water, flowing out of a rock at about half ebb of the tide, from which they filled their kegs. Three of the men chose to stay on the island, and take their chance for some vessel to take them off.

On the 27th of the 12th month, they left this island, and steered for Easter Island; but passed it far to the leeward. They then directed their course for Juan Fernandez, which was about twenty-five hundred miles E.S.E. from them. On the 10th of 1st month, 1821, Matthew P. Joy. the 2d mate, died, and his body was launched into the deep. His constitution was slender, and it was supposed that his sufferings, though great, were not the immediate cause of his death. On the 12th, the mate's boat separated from the other two, and did not fall in with them afterwards. The situation of the mate and his crew, became daily more and more distressing. The weather was mostly calm, the sun hot and scorching. They were growing weaker and weaker by want of food, and yet, such was their distance from land, they were obliged to lessen their allowance nearly one half. On the 20th, a black man died. On the 28th, they found, on calculation, that their allowance, only one and a half ounce of bread per day to a man, would be exhausted in fourteen days; and that this allowance was not sufficient to sustain life. They therefore determined to extend the indulgence, and take the consequence, whether to live or die. On the 8th of the 2d month, another of the crew died. From this time to the 17th, their sufferings were extreme. At 7 o'clock A.M. of that day, they were aroused from a lethargy by the cheering cry of the steersman, "there's a sail." The boat was soon descried by the vessel, the brig Indian, Captain Grozier, of London, which shortened sail and took them on board, lat. 33° 45' S. lon. 81° 3' W. They were treated by Captain Grozier with all the care and tenderness, which their weal condition required. On the same day they made Massafuero [*sic*], and on the 25th arrived at Valparaiso.

Captain Pollard and Charles Ramsdell, the only survivors in the captain's boat, were taken up on the 23d of the 2d month, 1821, by the ship Dauphin, of Nantucket, Captain Zimri Coffin, in lat. 37° S. off St. Mary's. The captain relates, that, after the mate's boat was separated from the others, they made what progress their weak condition would permit, towards the island of Juan Fernandez; but contrary winds and calm weather, together with the extreme debility of the crew, prevented their making much progress. On the 29th of the 1st month, the 2d mate's boat separated from the captain's in the night, at which time their provisions were wholly exhausted, since which they have not been heard from.

We shall not attempt a sketch of the sufferings of the crews of these boats. Imagination may picture the horrors of their situation, and the extremes to which they were driven to sustain life, but no power of the imagination can heighten the dreadful reality.

The following is an account of the whole crew.

In the captain's boat two survived, viz: Captain Pollard and Charles Ramsdell. In the mate's boat three survived, viz: Owen Chase, the mate, Benjamin Lawrence, and Thomas Nickerson. Left on Ducie's Island and afterwards taken off, Seth Weeks, William Wright, and Thomas Chapple. One left the ship before the accident. In the second mate's boat, when separated from the captain's, three. Dead nine, which added to the second mate's crew, doubtless lost, makes total deaths twelve.

From Henry T. Cheever, *The Whale and His Captors; or, The Whaleman's Adventures, and the Whale's Biography*

(New York: Harper & Brothers, 1850)

INTRODUCTION

Henry Theodore Cheever (1814–1897) was the editor of the *New York Evangelist* from 1849 to 1852. In the early 1840s, he voyaged as a convalescent on the whale-ship *Commodore Preble*: in late 1849 or early 1850, he published *The Whale and His Captors.* In between, Cheever had scathingly reviewed *Typee,* taking offense at Melville's account of missionaries (hardly surprising) and concluding the book "cannot escape severe condemnation."

Melville seems to have missed this work when he returned from England in the winter of 1850. There is little reason to think Cheever would have escaped the same kind of ridicule—perhaps bitterer—that Melville heaped on William Scoresby had he known from the start that he was competing with his former reviewer (the book was instantly reprinted in Britain as *The Whaleman's Adventures* with an introduction by Scoresby, Melville's "Captain Sleet").

Howard Vincent, in *The Trying-Out of* Moby-Dick, points out that when Melville turned to Cheever late in the composition of *Moby-Dick,* he did so for detail, not structure or inspiration. Nevertheless, Cheever's book is exactly the kind of book Melville proposed to Bentley—wild legends of the sperm whale fishery—but all packaged as an evangelical tract. But perhaps the most important aspect of Cheever's book is that although he was at sea at the same time as Melville, Cheever's whaling world appears incrementally more modern—more up-to-date—than Melville's. The difference is just sufficient that it might behoove us to read *Moby-Dick*, at least to some extent, as a historical novel.

Despite Vincent's assessment of provable borrowings from Cheever, an attentive reader might be struck by the number of analogous passages in the two books, and Melville's relationship to Cheever surely deserves further study.

Cheever had Beale's essay on the sperm whale but not his book. He depended heavily on Scoresby's *Arctic Regions*, from which, for instance, came the anecdote of the *Royal Bounty.* Cheever's retelling is perhaps even more thrilling than his original. Bercaw #136.

PRIMARY SOURCE

The substance called ambergris, and highly prized in perfumery, is obtained from the sperm whale, being formed, it is thought, in that state of the system which calls for a cathartic. A peck of Morrison's or Brandreth's pills, or the homeopathic dose of a pound of calomel and jalap, would probably remove obstructions in the creature's abdominal viscera, and prevent the formation of ambergris concretions, with undoubted benefit to the whale's corporation from the drastic operation, though it might be a loss to the perfumer and the Asiatic gastronomer, inasmuch as we learn from the Materia Medica that in Asia and parts of Africa ambergris is not only used as a medicine and a perfume, but considerable use also is made of it in cooking, by adding it to several dishes as a spice. A great quantity of it also is constantly brought by the pilgrims who travel to Mecca, probably to offer it there in fumigations, like as frankincense in the worship of the Church of Rome. . . .

A few years since, one of the most active and energetic of our whaling captains was thus taken overboard by the line, and had the singular good fortune to survive to tell the story. The whale was sounding very swiftly when the line became entangled. The boat-steerer, who was at his post in the stern of the boat, tending the line, instantly threw the turn off the loggerhead, and the tangled part ran forward and caught in the bow. The captain was seen to stoop to clear it, and then at once disappeared. The boat-steerer seized the hatchet, which is always at hand, and chopped the line, with the faint hope that when it slackened the captain could extricate himself.

The accident being so sudden and dreadful as almost to stupify [*sic*] the amazed crew, neither of them spake a word, but each eye was fixed upon the sea with fearful interest. Several minutes had elapsed, and the last hope was expiring, when an object was seen to rise to the surface a short ways from the boat, which, though exhibiting no sign of animation, was speedily reached, and the body of the captain, apparently lifeless, was lifted into the boat. It was evident that vitality was not extinct, and, to the joy of the little crew, symptoms of consciousness became visible in a few minutes, and the oars were lustily plied to reach the ship. By means of the usual remedies, the resuscitated captain was in a few days, in his own words, "as good as new."

In giving an account of the accident and his singular escape, he said that, as soon as he discovered the line had caught in the bow of the boat, he stooped to clear it, and attempted to throw it out from the "chock," so that it might run free. In doing this he must have caught a turn round his left wrist, and felt himself dragged overboard. He was perfectly conscious while he was rushing down, down, with unknown force and swiftness; and it appeared to him that his arm would be torn from his body, so great was the resistance of the water. He was well aware of his perilous condition, and that his only chance for life was to cut the line. But he could not remove his right arm from his side, to which it was pressed by the force of the element through which he was drawn.

When he first opened his eyes, it appeared as if a stream of fire was passing before them; but as he descended it grew dark, and he felt a terrible pressure on his brain, and a roaring as of thunder in his ears. Yet he was conscious of his situation, and made several efforts to reach the knife that was in his belt. At last, as he felt his strength failing and his brain reeling, the line for an instant slackened; he reached his knife, and instantly that the line again became taut, its edge was upon it, and by a desperate effort of his exhausted energies he freed himself. After this he only remembered a feeling of suffocation, a gurgling spasm, and all was over, until he awoke to an agonizing sense of pain in the boat. . . .

What goes under the name of the sperm whale's head is nearly one third of the monster's length. It is customary to sever this entirely at first, and let it tow astern, while the rest of the carcass is being stripped of its very valuable blubber. Utterly unlike the right whale, which has no teeth, the head of a sperm whale is square, the end of it something like the largest mill-log ever seen, sawed off straight. It is truly a prodigious mass of organized, compact matter, with which this animal has been known sometimes to butt like a ram against the sides of a ship and break it to pieces. They often go in this way "head on" to boats, but are generally pricked off and turned with a lance, or dexterously avoided.

From what may be called the top of the forehead to the roof of the mouth of this square-faced *sui generis* monster, it measured nine feet in a straight line, and there was a corresponding breadth and depth of forehead, so that, with its prodigious volume of brain (head matter), and so large a facial angle, the bust of this creature is most favorably commended to the fingers of phrenologists. Is it not a little surprising, that in the researches of comparative phrenology the cranium of the great sperm whale should be overlooked?

For the matter of room, a phrenologist might keep shop in it, and light it up, if he chose, with its own brains, and there point out to visitors by the lamp-light the places in the walls and ceiling where the different organs lay. It would be like a painter at Rome who should open his studio on the Parthenon; the celestial gods would be eying him from the ceiling; deified men and the infernals would be looking on him from all around. And if the aforesaid phrenologist and the favored artist should not alike become masters, under circumstances so imposing, it would be nobody's fault but their own. . . .

Chapter XIV

Remarkable Events in the Annals of Whaling

O'er the deep! o'er the deep!
Where the whale, and the shark, and the sword-fish sleep:
On the craggy ice, in the frozen air,
Heedless of dangers if game be but there,
Encountering all the great whale to snare.

—*Anon.*

The prodigious speed and strength of the gigantic whale, and the resulting danger to his captors referred to, in general terms, in the last chapter, are practically illustrated by two remarkable incidents, occurring, the one to the English, and the other in the American whale fishery.

On the 28th of May, 1817, the Royal Bounty, an English ship, fell in with a great number of whales in seventy degrees twenty-five minutes north latitude, and longitude five degrees east. There was neither ice nor land in sight. The boats were manned and sent in pursuit, and after a chase of five hours, one of them, which had rowed out of sight of the ship, struck one of the whales. This was about four o'clock in the morning. The captain directed the course of the ship to the point where he had last seen the boats, and about eight o'clock got sight of one, which displayed the signal of being fast. Soon after, another boat approached the first, and struck a second harpoon; and by mid-day two more harpoons were made fast.

But such was the astonishing vigor of this whale, that although it constantly dragged through the water from four to six boats, together with sixteen hundred fathoms of line, yet it pursued its flight nearly as fast as a boat could row, and whenever one passed beyond its tail it would dive. All endeavors to lance it were therefore vain, and the crews of the loose boats moored to those that were fast, the whale all the time steadily towing them on.

At eight o'clock in the evening a line was taken to the ship, with a view of retarding its flight, and topsails were lowered, but the harpoon drew. In three hours another line was taken on board, which immediately snapped. At four in the afternoon of the next day, thirty-six hours after the whale was first struck, two of the fast lines were taken on board the ship.

The wind blowing a moderately brisk breeze, the top-gallant-sails were taken in, the courses hauled up, and the top-sails were taken in, the course hauled up, and the top-sails clewed down; yet in this situation she was towed directly to windward, for an hour and a half, with the velocity of one and a half to two knots an hours, the whale all the while beating the water with its fins and tail, so that the sea around was in a continual foam. At length, near eight o'clock, after forty hours of incessant exertion, this tenacious assertor of his vast animal vigor and territorial rights was killed.

There is an instance given by Captain Scoresby where a Greenland whale was at last killed who had drawn out ten thousand four hundred and forty yards, or about six miles of line, attached to fifteen harpoons, besides taking one boat entirely under water, which disappeared and was never seen, the harpoons by which it was held to the whale probably drawing out under the immense pressure, and leaving it to sink, or to rise under the ice.

But the most dreadful display of the whale's strength and prowess yet authentically recorded was that made upon the American whale ship Essex, Captain Pollard, which sailed from Nantucket for the Pacific Ocean in August, 1819. Late in the fall of the same year, when in latitude forty of the South Pacific, a school of sperm whales were discovered, and three boats were manned and sent in pursuit. The mate's boat was struck by one of them, and he was obliged to return to the ship in order to repair the damage.

While he was engaged in that work, a sperm whale, judged to be eighty-five feet long, broke water about twenty rods from the ship, on her weather bow. He was going at the rate of about three knots an hour, and the ship at nearly the same rate, when he struck the bows of the vessel just forward of her chains.

At the shock produced by the collision of two such mighty masses of matter in motion, the ship shook like a leaf. The seemingly malicious whale dove and passed under the ship, grazing her keel, and then appeared at about the distance of a ship's length, lashing the sea with fins and tail, as if suffering the most horrible agony. He was evidently hurt by the collision, and blindly frantic with instinctive rage.

In a few minutes he seemed to recover himself, and started with great speed directly across the vessel's course to the windward. Mean time the hands on board discovered the ship to be gradually settling down at the bows, and the pumps were to be rigged. While working at them, one of the men cried out, "God have mercy! he comes again!"

The whale had turned at about one hundred rods from the ship, and was making for her with double his former speed, his pathway white with foam. Rushing head on, he struck her again at the bow, and the tremendous blow stove her in. The whale dived under again and disappeared, and the ship foundered in ten minutes from the first collision.

After incredible hardships and sufferings in their open boats on the 20th of December the survivors of the catastrophe reached the low island called Ducies, in latitude twenty-four degrees forty minutes south, longitude one hundred and twenty-four degrees forty minutes west. It was a mere sand-bank, nearly barren, which supplied them only with water and sea-fowl. On this uninhabited island, dreary as it was, three of the men chose to remain, remain than again commit themselves to the uncertainties of the sea. They have never since been heard from, the island being seldom visited.

On the 27th of December, the three boats, with the remainder of the men, put away together for the Island of Juan Fernandez, at a distance of two thousand miles. The mate's boat was taken up by the Indian, of London, on the 19th of February, ninety-three days from the time of the catastrophe, with only three living survivors.

The captain's boat was fallen in with by the Dauphin, of Nantucket, on the 23d of the same month, having only two men living, whose lives had been eked out only through that last resort of hunger in the wretched, which words shudder to relate! Out of a crew of twenty, five only survived to make the ear of the world tingle at their strange, eventful story.

There is one other instance of the immediate shipwreck of a whaler by the shock of those mighty leviathans, that of the Union, of Nantucket, Captain Gardner, which was totally lost, in the year 1807, between Nantucket and the Azores, by a similar concussion. A merchant brig, plying between Panama and one of the ports of Western Mexico, has lately met with the same disaster, but without loss of life, the passengers and crew being all rescued by an American whale ship.

Another form of the perils of whaling is illustrated in the following incidents, taken from an authentic communication in one of the religious newspapers of the day, which we insert here in order to complete this Daguerreotype Gallery of Life and Adventures in a Whale Ship.

A few years ago, the captain of a whale ship was on a cruise in the Pacific Ocean. There were three boats attached to the ship. Early one morning a whale appeared. Two boats were sent to capture it. They fastened to the whale, and were soon drawn by this monster of the deep out of sight of the ship. An hour or two passed along, when, suddenly, another whale rose in the water, but a few rods from the vessel. The temptation to attempt its capture was too strong to be resisted. The captain ordered the remaining boat to be lowered, and, leaving but one man and two boys to take care of the ship, sprang into the boat with the rest of the crew.

Soon the harpoon was plunged into the whale, and they were carried, with almost the speed of the wind, about fifteen miles from the ship. Then the whale plunged perpendicularly down into the depths of the ocean. It was not long ere they saw him, fathoms deep in the crystal waters, rushing up, with open jaws, to destroy the boat. By skillfully sheering the boat, the whale missed his aim, and, thrusting his mammoth head some fifteen or twenty feet into the air, he fell over upon his side, and again disappeared in the fathomless sea. Soon he reappeared in the almost transparent abyss, again rushing upwards to attack the boat. Again he was failed.

The third time he descended, and as he arose, with invigorated fury, he struck the boat in the center of the keel, threw it some fifteen feet into the air, and, scattering the crew and fragments of the boat over the waves, again plunged into the deep and disappeared. The captain and the crew were now in the water, clinging to the pieces of the demolished boat. They were fifteen miles from the ship, and could not be seen from its deck. The other boats were gone, they knew not where. Apparently, every chance of rescue was cut off, and nothing awaited them but a watery grave. It was twelve o'clock at noon. The hours of one, two, three, four, five, and six passed slowly away, and still they were floating, almost exhausted, upon the heaving billows of the Pacific. When the ship arose on the swelling seas, they could just catch a glimpse of her rolling spars.

"Oh! how fervently I prayed," said one of these mariners, when afterward relating the scene, "that God would in some way providentially interpose and save our lives! I thought of my wife, of my little children, of my prayerless life, of the awful account I had to render at the bar of God for grieving the Spirit and neglecting the Savior. All the horrors of this dreadful death were forgotten in the thought, that in one short hour I was to render up an account to God for years of ingratitude and disobedience. Oh! thought I, if I were only a Christian, what a solace would it be to me as I sink into this watery grave!"

The sun had now disappeared behind the distant waves, and the darkening shades of a dreary night were settling down over the ocean. Just then they descried, dim in the dusky distance, one of the absent boats returning to the ship. It was, however, far off, apparently beyond the reach of their loudest outcries. Impelled by the energies of despair, they simultaneously raised a shout, which blended with the wash of the waves and sighing of the breeze, and the boat continued on its way. Again they raised another shout, and it was also unavailing.

The shades of the night were deepening, the boat rapidly passing by them. Almost frenzied at their terrible condition, they raised another cry. The sound of that distant shriek fell faintly upon the ears of the boatmen, and they rested on their oars. Another shout, which almost lacerated their throats, was raised, and the boat turned in pursuit. They were taken from the water, and carried almost lifeless to the ship.

In another authentic instance, when a boat was chasing a whale, he suddenly turned to windward, and made directly for his pursuers, who were so excited by the chase as to be blind to danger. On, therefore, they madly rushed, without trying to avoid the infuriated monster, so eager were they to plunge into him their irons, till the boat struck with such force upon the whale's head as to throw the oarsmen from their thwarts. At the same moment, the boatsteerer let fly his two harpoons into the mammoth body, which rolled over on its back; and before the boat could get clear of danger, being to the windward, a heavy sea struck it and threw them directly into the whale's mouth! All, of course, sprang for their lives, and they had barely time to throw themselves clear of the boat before it was crushed to pieces by those ponderous jaws, and its ejected crew were providentially all picked up by another boat.

Such are the dangers which are continually incurred in the whale fishery, equal almost to those of the field of battle. We often wonder that so many escape with their lives from a battle field; and we equally wonder that, comparatively, so few perish in this most hazardous pursuit. A boat, almost as frail as a bubble, approaches the side of a whale, slumbering upon the ocean, sixty or eighty feet in length, and a harpoon is plunged into his body. His efforts to destroy his tormentors or escape from them, as we have again and again learned, are terrific. The ocean is lashed into foam by blows from his enormous flukes, which would almost dash in the ribs of a man-of-war. Often he rushes at the boat with lightning speed and with open jaws, and it is crushed like an egg-shell in its mouth.

In this frightful warfare many are maimed, and many lives are annually lost. But some whales are worth between two and three thousand dollars, and this is majestic game to hunt. He, however, who earns his bread through the perils and hardships of this pursuit, has truly a hard lot in life. He is but a transient visitor at his home. Amid the solitude of the ocean he passes the greater portion of his days; and if he survives the perils of his adventurous pursuit, the storms of the ocean, and the pestilence of different climes, he usually finds that the friends of his youth are all gone, and that he is almost a stranger at his own fireside. And yet this mode of his life has its own joys and emoluments, for, if ordinarily successful, in the course of fifteen or twenty years a whaleman will lay up a moderate competence for the rest of his days, and meanwhile, notwithstanding the unfavorable influences which are often at work in the whale ship, many are forming noble characters.

Although it is no genial soil, yet virtue, humanity, true nobility, and the fear of God, *can* live and grow in a whale ship, both fore and aft. I have met them on this present voyage, and in some signal instances elsewhere, which it would be base ingratitude and a denying of God's grace not to acknowledge and give credit for. But who that knows it as I do, would choose a life in a whale ship, or life any where at sea! Who does not rather say, with one that knew whereof he spake,

> Eternal ocean! old majestic sea!
> Ever I love from shore to look on thee,
> And sometimes on thy billowy back to ride,
> And sometimes o'er thy summer breast to glide;
> But let me *live* on land, where rivers run,
> Where shady trees may screen me from the sun
> Where I may feel, secure, the fragrant air;
> Where, whate'er toil or wearying pains I bear,
> Those Eyes which look away all human ill,
> May shed on me their still, sweet constant light,
> And the hearts I love may, day and night,
> Be found beside me safe and clustering still.

HERMAN MELVILLE, *CLAREL: A POEM AND PILGRIMAGE TO THE HOLY LAND*

(1876), 1.37.6–108

INTRODUCTION

After publishing *Battle-Pieces or Aspects of the War*, his collection of Civil War poems, in 1866, Melville continued to study the history and practice of poetry. In the late 1860s, he especially turned to the work of Matthew Arnold (1822–1888), whose "Empedocles on Etna" could almost serve as a prospectus to Melville's own *Clarel: a Poem and a Pilgrimage in the Holy Land* (1876). Melville's long poem of faith and doubt gave him an opportunity to revisit the captain and mate of the *Essex,* employing a physical description of the man he mistakenly thought was Owen Chase along with the repeated misfortunes (here adapted and heightened) of Captain George Pollard. The philosophizing in the excerpt from the poem below is more reminiscent of Melville and his friend George J. Adler (1821–1868) on their autumn 1849 passage to England in the *St. Lawrence* than of anything known about the internal mental strivings of Chase or Pollard. Adler himself had bouts of insanity, leading to his death while incarcerated in an asylum. In the tradition of the elegy, Adler would not be named even if the following passage were meant to recall him.

It may have been closer to the time of writing *Clarel* than of *Moby-Dick* that Melville added to his copy of Chase, in blue-green crayon, the following note: "Since writing the foregoing I—somewhere about 1850–3—saw Capt. Pollard on the island of Nantucket, and exchanged some words with him. To the islanders he was a nobody—to me, the most impressive man, tho' wholly unassuming, even humble—that I ever encountered."

PRIMARY SOURCE

A Sketch

For ease upon the ground they sit;
And Rolfe, with eye still following
Where Nehemiah slow footed it,
Asked Clarel: "Know you anything
Of this man's prior life at all?"
"Nothing," said Clarel.—"I recall,"
Said Rolfe, "a mariner like him."
"A mariner?"—"Yes; one whom grim
Disaster made as meek as he
There plodding." Vine here showed the zest
Of a deep human interest:
"We crave of you his history."
And Rolfe began: "Scarce would I tell
Of what this mariner befell—
So much is it with cloud o'ercast—
Were he not now gone home at last
Into the green land of the dead,
Where he encamps and peace is shed.
Hardy he was, sanguine and bold,
The master of a ship. His mind
In night-watch frequent he unrolled—
As seamen sometimes are inclined—
On serious topics, to his mate,
A man to creed austere resigned.
The master ever spurned at fate,
Calvin's or Zeno's. Always still
Man-like he stood by man's free will
And power to effect each thing he would,
Did reason but pronounce it good.
The subaltern held in humble way
That still heaven's over-rulings sway
Will and event.
"On waters far,
Where map-man never made survey,
Gliding along in easy plight,
The strong one brake the lull of night
Emphatic in his willful war—
But staggered, for there came a jar
With fell arrest to keel and speech:
A hidden rock. The pound—the grind—

Collapsing sails o'er deck declined—
Sleek billows curling in the breach,
And nature with her neutral mind.
A wreck. 'T was in the former days,
Those waters then obscure; a maze;
The isles were dreaded—every chain;
Better to brave the immense of sea,
And venture for the Spanish Main,
Beating and rowing against the trades,
Than float to valleys 'neath the lee,
Nor far removed, and palmy shades.
So deemed he, strongly erring there.
To boats they take; the weather fair—
Never the sky a cloudlet knew;
A temperate wind unvarying blew
Week after week; yet came despair;
The bread tho' doled, and water stored.
Ran low and lower—ceased. They burn—
They agonize till crime abhorred
Lawful might be. O trade-wind, turn!
  "Well may some items sleep unrolled—
Never by the one survivor told.
Him they picked up, where, cuddled down,
They saw the jacketed skeleton,
Lone in the only boat that lived—
His signal frittered to a shred.
  "'Strong need'st thou be,' the rescuers said,
'Who has such trial sole survived.'
'I willed it,' gasped he. And the man,
Renewed ashore, pushed off again.
How bravely sailed the pennoned ship
Bound outward on her sealing trip
Antarctic. Yes; but who returns
Too soon, regaining port by land
Who left it by the bay? What spurns
Were his that so could countermand?
Nor mutineer, nor rock, nor gale
Nor leak had foiled him. No; a whale
Of purpose aiming, stove the bow:
They foundered. To the master now
Owners and neighbors all impute
An inauspiciousness. His wife—
Gentle, but unheroic—she,

Poor thing, at heart knew bitter strife
Between her love and her simplicity:
A Jonah is he?—And men bruit
The story. None will give him place
In a third venture. Came the day
Dire need constrained the man to pace
A night patrolman on the quay
Watching the bales till morning hour
Through fair and foul. Never he smiled;
Call him, and he would come; not sour
In spirit, but meek and reconciled;
Patient he was, he none withstood;
Oft on some secret thing would brood.
He ate what came, though but a crust;
In Calvin's creed he put his trust;
Praised heaven, and said that God was good,
And his calamity but just.
So Silvio Pellico from cell-door
Forth tottering, after dungeoned years,
Crippled and bleached, and dead his peers:
'Grateful, I thank the Emperor.'"

HERMAN MELVILLE'S SEA ROUTES

It is perhaps useful to have a list of *all* Melville's voyages, both whaling and otherwise. The name of each vessel is followed by a list of locations and rough dates that will help to locate Melville's position. Where only coordinates of latitude and longitude are given, Melville's vessel was in the open sea, far from land. The different tracks are derived from a variety of sources, but most can be found in Parker (1996, 2002). Dates are approximate, as there might be different dates for landfall and actual disembarking, and Melville often misdated his journals. Some excursions in smaller vessels, too, especially in Scotland and the Levant (eastern Mediterranean) are unreported here. It is curious to note that if the data are complete, once the *Acushnet* headed south to the Bahama Banks, *there is no demonstrable overlap with any of Melville's actual voyages and the fictional voyage of the* Pequod.

I. MELVILLE'S PASSAGES AS A SAILOR

St. Lawrence **(Packet Ship)**

New York to Liverpool early June to early July 1839
Liverpool to New York mid-August to late September 1839

Acushnet (Whaler)

New Bedford bound for the Pacific Ocean early January 1841
Bahama Banks late winter 1841
Rio de Janeiro mid-March 1841
Cape Horn mid-April 1841
South of Juan Fernandez early May 1841
Santa Harbor (Peru) late June 1841
Roughly 5° S latitude, 105° W longitude late July to mid-October 1841
Albemarle Island (Galapagos) late October 1841
Galapagos early to mid-November 1841
Tombez (Peru) early December 1841
Off Ecuador late December 1841
Galapagos early January 1842
5° 31'S 102° 41'W mid-February 1842
4° 47'S 105° 54'W early March 1842

The *Acushnet*'s closest point of approach to the location of the *Essex* sinking (roughly 0° Latitude, 119° West Longitude) falls between these two entries. The *Acushnet* passed within about 300 land miles of the site—not a great distance given the vastness of the Pacific Ocean.

4° 09'S 122° 54'W mid-April 1842
0°12'S 127° 52'W early May 1842
Marquesas late June 1842

Lucy Ann (Whaler)

Marquesas early August 1842
Tahiti late September 1842

Charles and Henry (Whaler)

Tahiti early November 1842
Two days east of 34° 10'S 87° 40'W late January 1843 (off Santiago, Chile, and heading toward Masafuero)
14° S 113° 42'W early February 1843 (halfway on a line between Pitcairn and the Galapagos)
Lahaina Maui early May 1843

In making its way from this last reported position to Hawaii, the *Charles and Henry* may also have passed close to the site of the *Essex* sinking.

United States (Frigate)

Lahaina mid-August 1843
Nukahiva early October 1843
Tahiti mid-October 1843
Juan Fernandez late October 1843
Valparaiso late November 1843
Callao January to February 1844
Mazatlan late March to mid-April 1844
Callao early June to early July 1844
Rio mid-August 1844
Cape Cod early October 1844
Boston early October 1844

II. MELVILLE'S PASSAGES AS A PASSENGER

Southampton (Packet Ship)

New York to London early October to early November 1849

Independence (Packet Ship)

Portsmouth to New York late December 1849 to late January 1850

Glasgow (Screw Steamer)

New York to Glasgow dep. mid-October 1856

Egyptian (Steamer)

Leave Liverpool mid-November
Gibraltar late November 1856
Malta late November 1856
Syra (Greek Islands) early December 1856
Salonica early December 1856
Hellespont mid-December 1856
Constantinople mid-December 1856

Acadia (Steamer)

Constantinople to Smyrna mid to late December 1856
Smyrna to Syra late December 1856
Syra to Alexandria late December 1856

Unknown (Steamer)

Alexandria to Jaffa early January 1857

***Acquila Imperiale* (Austrian Steamer)**

Jaffa to Beirut late January 1857

***Smirne* (Austrian Steamer)**

Beirut to Smyrna early February 1857

***Italia* (Steamer)**

Smyrna to Piraeus early February 1857

***Cydnus* (French Steamer)**

Piraeus to Messina mid-February 1857

Unknown (Neapolitan Steamer)

Messina to Naples mid-February 1857

***Aventime* (French Steamer)**

Civita Vecchia to Leghorn late March 1857

***City of Manchester* (Steamer)**

Liverpool to New York early May to late May 1857

***Meteor* (Clipper Ship)**

New York late May 1860
Crossed Tropic late June 1860
Crossed Line late June 1860
18° 30'S 33° 30'W early July 1860
43° S 49° W late July 1860
Staten Land early August 1860
Cape Horn early August 1860
19° S early September
5° N mid-September 1860
15° N S late September 1860
San Francisco mid-October 1860

Unknown (Steamer)

San Francisco to Panama leave late October 1860, arrive 13 days later

North Star (Steamer)

Arrive Manhattan from Panama mid-November 1860 after 10-day passage

Orinoco (Passenger Steamer)

New York to Bermuda arrive mid-March 1888

Trinidad (Passenger Steamer)

Bermuda to New York depart late March 1888

INDEX

ABOUT THE AUTHOR

R.D. (Robert Durwood) Madison is professor emeritus of English, U.S. Naval Academy. He is also associate general editor of the Northwestern-Newberry Edition of the Writings of Herman Melville, textual editor and board member of The Writings of James Fenimore Cooper, board member of *LEAR: Literature in the Early American Republic*, and board member and book-review editor for *Nautilus: A Maritime Journal of Literature, History, and Culture*. He has edited more than a dozen critical editions for both scholarly and commercial presses, including *The* Bounty *Mutiny*, Jack London's *The Cruise of the Snark,* and R. D. Blackmore's *Lorna Doone.*

www.ingramcontent.com/pod-product-compliance
Ingram Content Group UK Ltd.
Pitfield, Milton Keynes, MK11 3LW, UK
UKHW020219041026
15470UKWH00012B/665

9 781440 850073